Acquisition and Reproduction of Color Images:

Colorimetric and Multispectral Approaches

by

Jon Y. Hardeberg, Ph.D.

Dissertation.com • USA • 2001

Acquisition and Reproduction of Color Images:
Colorimetric and Multispectral Approaches

Copyright © 1999–2001 Jon Hardeberg
All rights reserved.

Dissertation.com
USA • 2001

ISBN: 1-58112-135-0

additional information and color ebook edition at:
www.dissertation.com/library/1121350a.htm

Preface

In the last two decades we have seen the field of digital color imaging emerging from specialized scientific applications, into being a part of the daily lives of most people in industrialized countries. Broadcast television, computers, newspapers and magazines, are just a few examples of technologies relying on digital color imaging.

The increased use of color has brought with it new challenges and problems. Digital color imaging is a research field with great prospects, since many problems are still unsolved.

One typical case that illustrates the problems that needs to be solved is that of an e-commerce business. Prospective customers use the World Wide Web to evaluate their products, typically by visualizing them on the computer monitor or by printing the product images on a desktop printer. If the customer decides to buy, and the color of the delivered product is not what she expected, she might decide to return it.

The book you're holding in your hands (or maybe you're reading it on your computer or eBook device) is the second edition of my Ph.D. dissertation. It is my hope that this book may be of service to the academic and industrial color imaging community, in many ways:

- As support for learning and teaching.

- Through the advances in the state-of-the-art represented by the proposed methods for the acquisition and reproduction of high quality digital color images.

- By advocating the multispectral approach to color imaging, that is to use more than three color channels, in order to overcome the problems of metamerism.

■ By giving ideas for further research concerning the technology, science, and art of color imaging.

Finally, I have to apologize that there's no color in the printed edition of this book. I find it amazing that color printing of a book like this is still not found to be cost-effective, even using an on-demand printing model. The only consolation in this is that I don't expect to be out of work for many many years to come!

Redmond, Washington, USA, June 2001

Abstract

The goal of the work reported in this dissertation is to develop methods for the acquisition and reproduction of high quality digital color images. To reach this goal it is necessary to understand and control the way in which the different devices involved in the entire color imaging chain treat colors. Therefore we addressed the problem of *colorimetric characterization* of scanners and printers, providing efficient and colorimetrically accurate means of conversion between a device-independent color space such as the CIELAB space, and the device-dependent color spaces of a scanner and a printer.

First, we propose a new method for the colorimetric characterization of color scanners. It consists of applying a non-linear correction to the scanner RGB values followed by a 3rd order 3D polynomial regression function directly to CIELAB space. This method gives very good results in terms of residual color differences. The method has been successfully applied to several color image acquisition devices, including digital cameras. Together with other proposed algorithms for image quality enhancements it has allowed us to obtain very high quality digital color images of fine art paintings.

An original method for the colorimetric characterization of a printer is then proposed. The method is based on a computational geometry approach. It uses a 3D triangulation technique to build a tetrahedral partition of the printer color gamut volume and it generates a surrounding structure enclosing the definition domain. The characterization provides the inverse transformation from the device-independent color space CIELAB to the device-dependent color space CMY, taking into account both colorimetric properties of the printer, and color gamut mapping.

To further improve the color precision and color fidelity we have performed another study concerning the acquisition of multispectral images using a monochrome digital camera together with a set of $K > 3$ carefully selected color filters. Several important issues are addressed in this study. A first step is to perform a spectral characterization of the image acquisition system to establish the spectral model. The choice of color chart for this characterization is found to be very important, and a new method for the design of an optimized color chart is proposed. Several methods for an optimized selection of color filters are then proposed, based on the spectral properties of the camera, the illuminant, and a set of color patches representative for the given application. To convert the camera output signals to device-independent data, several approaches are proposed and tested. One consists of applying regression methods to convert to a color space such as CIEXYZ or CIELAB. Another method is based on the spectral model of the acquisition system. By inverting the model, we can estimate the spectral reflectance of each pixel of the imaged surface. Finally we present an application where the acquired multispectral images are used to predict changes in color due to changes in the viewing illuminant. This method of illuminant simulation is found to be very accurate, and it works well on a wide range of illuminants having very different spectral properties. The proposed methods are evaluated by their theoretical properties, by simulations, and by experiments with a multispectral image acquisition system assembled using a CCD camera and a tunable filter in which the spectral transmittance can be controlled electronically.

Key words

Color imaging, colorimetry, colorimetric characterization, multispectral imaging, spectral characterization, filter selection, spectral reconstruction.

Acknowledgements

The work described in this document has been carried out at the Signal and Image Processing Department at the Ecole Nationale Supérieure des

Télécommunications (ENST) in Paris, France. This endeavor would not have been completed without the support and help of many people. You all deserve thanks! Running the risk of forgetting someone important, I will mention some of you in particular.

First of all, I would like to thank my advisors Francis Schmitt and Hans Brettel. Thanks for showing me the way into the wonderful world of color, and for guiding and encouraging me throughout my tenure at the ENST. A big thank you goes then to all my colleagues and friends at the ENST. To the former and present Ph.D. students with whom I have shared moments of work and pleasure: Anne, Bert, Dimitri, Florence, Geneviève, Jean-Pierre, Jorge, Lars, Maria-Elena, Mehdi, Raouf, Selim, Sophie, Sudha, Wirawan, Yann, and many others. You have meant a great deal to me, and I sincerely hope to be able to stay in touch with you! To those of you who have worked with me on colorful projects: Bahman, Brice, Frédéric, Henri, Ingeborg, Jean-Pierre. To our relations in the industry, for showing me that my research could be able to solve some real problems out there.

I thank the members of the jury, Jean-Marc Chassery, Jean-François Le Nest, Khadi Bouatouch, and Roger D. Hersch, for your fruitful questions, comments and suggestions, and for honoring me by your presence at my defence. Furthermore, I would like to thank the organizing committees of the international conferences I have attended, for accepting the papers I submitted. Not only has this allowed me to present my work, and learn about the work of other researchers in the field of color imaging, it has given me the opportunity to meet and discuss with several inspiring individuals. I have seen that the world of color is a small one, after all.

Funding has been provided by the Norwegian Research Council (Norges Forskningsråd). Its financial support is of course gratefully acknowledged.

I am also very grateful to Dan and RanDair at Conexant Systems, Inc. for giving me a great opportunity to continue to work in this enchanting field.

Finally, I cannot imagine doing anything without you, Kristine. Thank you for giving me a reason to move to Paris, thank you for putting up with me through times of hard pressure, thank you for my two lovely children, Pauline and Samuel, and most of all, thank you for being you.

Kirkland, Washington, USA, November 1999

About the author

Jon Y. Hardeberg received his sivilingeniør (M.Sc.) degree in signal processing from the Norwegian Institute of Technology (Trondheim, Norway) in 1995. He received his Ph.D. from the Ecole Nationale Supérieure des Télécommunications (Paris, France) in February 1999. His Ph.D. research concerned color image acquisition and reproduction, with applications in facsimile, fine-art paintings, and multi-spectral imaging. He is currently employed with Conexant Systems, Inc., where he designs, implements, and evaluates color imaging system solutions for multifunction peripherals and other imaging devices and systems. His professional memberships include IS&T, SPIE, and ISCC.

About this book

Hardeberg, Jon Yngve, **Acquisition and Reproduction of Color Images: Colorimetric and Multispectral Approaches.** A dissertation submitted in partial fulfillment of the degree of "Docteur de l'Ecole Nationale Supérieure des Télécommunications", Paris, France, February 1999.

First edition published in 1999 by Ecole Nationale Supérieure des Télécommunications, 46, rue Barrault, F-75634 Paris Cedex 13, France, under the title "Acquisition et reproduction d'images couleur : approches colorimétrique et multispectrale," Publication ENST 99 E 021.

Second edition published in 2001 by Universal Publishers / dissertation.com, 7525 NW 61 Terrace, Suite 2603, Parkland, FL 33067-2421.

Changes from the first edition include layout, adding of a List of Tables and a List of Figures, correction of miscellaneous typographical errors, translation to American English (you know, *color* instead of *colour*, etc.), modifications to Figures 2.7 on page 21 and 3.11 on page 68, modifications to the numbering of Figures and Equations, and this preface.

Further information

More information about the described research can be found at the following locations:

- At http://color.hardeberg.com/

- By contacting the author by E-mail: jon.hardeberg@conexant.com, jon@hardeberg.com

- By contacting Prof. Francis Schmitt or Dr. Hans Brettel by E-mail: schmitt@tsi.enst.fr, brettel@tsi.enst.fr, or through the following address: Ecole Nationale Supérieure des Télécommunications, Département TSI, 46, rue Barrault, F-75634 Paris Cedex 13, France

Contents

List of Figures

List of Tables

Chapter 1

Introduction

1.1 Motivation

The use of color in imaging continues to grow at an ever-increasing pace. Every day, most people in the industrialized parts of the world are users of color images that come from a wide range of imaging devices; for example color photographs, magazines, and television at home, computers with color displays, and color printers in the office.

As long as the colors are found to be approximately as expected, people are generally happy with their images. However, with the increased use of color images, people's quality requirements also have increased considerably. Just a few years ago, a computer graphics system capable of producing 256 different colors was more than enough for most users, while today, most computers that are sold have *true color* capabilities, being able to produce 16.7 million[1] colors.

Furthermore, several professions have particular needs for high-quality color images. Artists are very concerned about colors in their works, and so are the art historians and curators studying their works. The printing, graphic arts, and photography industries have been concerned about color imaging for a long time. Most of the color imaging standards and equip-

---

[1]Note that this number represents only the number of different colors that can be specified to the monitor ($2^8 \cdot 2^8 \cdot 2^8 = 16777216$); the actual number of distinguishable resulting colors is much lower, approximately on the order of 1 million (Pointer and Attridge, 1998).

ment used today have their roots in these industries. But the past twenty years have seen the field of digital color imaging emerging from specialized scientific applications into the mainstream of computing. Color is also extremely important in several other fields, such as the textile and clothing industry, automotive industry, decoration and architecture.

Digital color imaging systems process electronic information from various sources: images may come from the Internet, a remote sensing device, a local scanner, etc. After processing, a document is usually compressed and transmitted to several places via a computer network for viewing, editing or printing. To achieve color consistency throughout such a widely distributed system, it is necessary to understand and control the way in which the different devices involved in the entire color imaging chain treat colors. Each scanner, monitor, printer, or other color imaging device, senses or displays color in a different, device-dependent, way. One approach to exchanging images between these devices is to calibrate each color image acquisition and reproduction device to a device-independent color space. The exchange of images can then be done in this color space, which should conform to international standards.

However, colors represent an important but nevertheless limited aspect of the objects that surround us. They correspond to the human perception of its surface under given light conditions. For the needs of, for example, an art curator wanting to control any changes or ageing of the materials in a fine arts painting, or a publisher wanting extra high-fidelity color reproduction, it becomes necessary to provide a more complete spectral analysis of the objects. This requires technology and devices capable of acquiring multispectral images. A multispectral image may also be used to reproduce an image of the object, as it would have appeared under a given illuminant.

In this research, we have investigated several of the aspects mentioned above. We have developed novel algorithms for the colorimetric characterization of scanners and printers providing efficient and colorimetrically accurate means of conversion between a device-independent color space such as CIELAB, and the device-dependent color spaces of a scanner and a printer. Furthermore, we have developed algorithms for multispectral image capture using a CCD camera with carefully selected optical filters. The developed algorithms have been used for several applications, such as fine-arts archiving and color facsimile.

1.2 Dissertation outline

This thesis is organized as follows. Chapter 2 provides an introduction to light, objects, human color vision, and the interaction between them, gives an introduction to important elements of colorimetry, and finally presents the subject of color imaging.

In Chapter 3, a methodology for the colorimetric characterization of color scanners is proposed. It consists of applying a non-linear correction to the scanner RGB values followed by a 3rd order 3D polynomial regression function directly to CIELAB space. This method gives very good results in terms of residual color differences. This is partly due to the fact that the RMS error that is minimized in the regression corresponds to ΔE_{ab}, which is well correlated to visual color differences. The method has been successfully applied to several color image acquisition devices.

In Chapter 4, various techniques for the digital acquisition and processing of high quality and high definition color images using a CCD camera are developed. The techniques have been applied to fine arts paintings on several occasions, e.g. for the making of a CDROM on the French painter Jean-Baptiste Camille Corot (1796-1876).

A novel method for the colorimetric characterization of a printer is proposed in Chapter 5. The method is based on a computational geometry approach. It uses a 3D triangulation technique to build a tetrahedral partition of the printer color gamut volume and it generates a surrounding structure enclosing the definition domain. The characterization provides the inverse transformation from the device-independent color space CIE-LAB to the device-dependent color space CMY, taking into account both colorimetric properties of the printer, and color gamut mapping.

We construct two 3D structures which provide us with a partition of the space into two sets of non-intersecting tetrahedra, an inner structure covering the printer gamut (i.e. the full set of the printable colors), and a surrounding structure, the union of these two structures covering the entire definition domain of the CIELAB space. These 3D structures allow us to easily determine if a CIELAB point is inside or outside the printer color gamut, to apply a gamut mapping technique when necessary, and then to compute by irregular tetrahedral interpolation the corresponding CMY values. We establish thus an empirical inverse printer model. This algorithm has been protected by a patent, and is now transferred to industry and used in commercial color management software.

In Chapter 6, we describe a system for the acquisition of multispectral images using a CCD camera with a set of optical filters. Several important issues are addressed in this study.

First, a spectral model of the acquisition system is established, and we propose methods to estimate its spectral sensitivities by capturing a color chart with patches of known spectral reflectance and by inverting the resulting system of linear equations. By simulations we evaluate the influence of acquisition noise on this process. The choice of color chart is found to be very important, and a method for the design of an optimized color chart is proposed.

We further discuss how the surface spectral reflectance of the imaged objects may be reconstructed from the camera responses. We perform a thorough statistical analysis of different databases of spectral reflectances, and we use the resulting statistical information along with the spectral properties of the camera and the illuminant to choose a set of optimal optical filters for a given application.

Finally we present an application where the acquired multispectral images are used to predict changes in color due to changes in the viewing illuminant. This method of illuminant simulation is found to be very accurate, and applicable to a wide range of illuminants having very different spectral properties.

In Chapter 7 the theoretical models and simulations of the previous chapter are validated in practice. An experimental multispectral camera was assembled using a professional monochrome CCD camera and an optical tunable filter. To be able to recover colorimetric and spectrophotometric information about the imaged surface from the camera output signals, two main approaches are proposed. One consists of applying an extended version of the colorimetric scanner characterization method described above to convert from the camera outputs to a device-independent color space such as CIEXYZ or CIELAB. Another method is based on the spectral model of the acquisition system. By inverting the model, we can estimate the spectral reflectance of each pixel of the imaged surface.

Finally, Chapter 8 concludes this dissertation and contains a discussion of possible future work based on the results reported here.

1.3 Notation used throughout this document

■ Vectors are represented in lowercase boldface letters, *e.g.* **a** and **θ**. They are generally written as column vectors,

$$\mathbf{a} = \begin{bmatrix} a_1 \\ a_2 \\ \vdots \\ a_N \end{bmatrix}.$$

■ Matrices are represented using uppercase boldface letters, *e.g.* **A** and **Θ**. The entry of matrix **A** in the ith line and the jth column is generally denoted a_{ij}. This may also be expressed as

$$\mathbf{A} = [a_{ij}] = \begin{bmatrix} a_{11} & a_{12} & \cdots & a_{1M} \\ a_{21} & a_{22} & \cdots & a_{2M} \\ \vdots & \vdots & \ddots & \vdots \\ a_{N1} & a_{N2} & \cdots & a_{NM} \end{bmatrix}.$$

■ An $(N \times M)$ matrix has N lines and M columns.

■ The transpose of a matrix is represented with a t in superscipt, *e.g.* $\mathbf{A}^t$.

■ The identity matrix of size $(N \times N)$ is denoted $\mathbf{I}_N$.

■ A vector space spanned by the P column vectors of a matrix $\mathbf{P} = [\mathbf{p}_1 \mathbf{p}_2 \ldots \mathbf{p}_P]$ is denoted the *range* of **P**, $R(\mathbf{P})$.

■ rank(**P**) is the dimension of $R(\mathbf{P})$.

■ Unless otherwise stated, the norm $\|\mathbf{x}\|$ of a N-vector is the 2-norm, defined as $\left(\sum_{i=1}^{N} x_i^2 \right)^{1/2}$.

■ The pseudoinverse is denoted by the $-$ sign in superscript, *e.g.* $\mathbf{A}^-$.

■ Note that a slightly different notation is used in Chapter 5 where uppercase boldface letters are used to denote vectors, and where the inner product of two vectors is denoted $\mathbf{A} \cdot \mathbf{B}$ instead of $\mathbf{a}^t \mathbf{b}$ which would be the case in the rest of the document.

Chapter 2

Color and imaging

THE INCREASED USE OF COLOR IMAGES HAS BROUGHT WITH IT NEW
CHALLENGES AND PROBLEMS. IN ORDER TO MEANINGFULLY RECORD
AND PROCESS COLOR IMAGES, IT IS ESSENTIAL TO UNDERSTAND THE
INTERACTION BETWEEN LIGHT, OBJECTS, AND HUMAN COLOR VISION
AND FURTHERMORE THE CAPABILITIES AND LIMITATIONS OF COLOR
IMAGING DEVICES. IN THIS CHAPTER WE PRESENT AN OVERVIEW OF
THESE BASICS, WITHOUT IN ANY WAY ASPIRING TO COMPETE WITH
COMPREHENSIVE TEXTBOOKS ON THESE SUBJECTS (LEGRAND, 1957,
KOWALISKI, 1990, SÈVE, 1996, WYSZECKI AND STILES, 1982, HUNT,
1995).

2.1 Introduction

What *is* color? This apparently simple question turns out to be rather difficult to answer concisely. The distinguished researcher Lars Sivik expresses
it as follows (Sivik, 1997).

> *Blessed are the "naive", those who do not know anything*
> *about color in a so-called scientific meaning — for them co-*
> *lor is no problem. Color is as self-evident as most other things*

*and phenomena in life, like night and day, up and down, air
and water. And all seeing humans know what color is. It
constitutes, together with form, our visual world. I have earlier used the analogy with St. Augustine's sentence about time:
"Everybody knows what time is — until you ask him to explain
what it is." It is the same with color.*

Misunderstandings are quite common when it comes to color. One reason is that the word color is given so many meanings — paint, CIE-values,
RGB-values, spectral radiation, perceptual sensations, color system notation, etc. In the following sections we will discuss some of the important
aspects of color and the relations between them. We establish a scientific
framework for the quantization of color. As a starting point we cite the
most widely accepted technical definition of color, given by the Committee on Colorimetry of the Optical Society of America in 1940, as cited in
Nimeroff (1972).

*Color consists of the characteristics of light other than spa-
tial and temporal inhomogeneities; light being that aspect of
radiant energy of which a human observer is aware through
the visual sensations which arise from the stimulation of the
retina of the eye.*

We see that this definition relates the *psychological* entities color and light
to the *physically* defined radiant energy in the part of the spectrum having a visual effect on the observer. The term *psychophysics* is thus often
employed in color science, meaning the science dealing with the relation
between the physical attributes of stimuli and the resulting sensations.

We will start this chapter by a presentation of the physical properties of
light and surfaces in Section 2.2. Colored light has varying radiant energy
for different wavelengths. Colored surfaces transmit and reflect different
amounts of the incident light for different wavelengths. The spectral interaction between light and surfaces represents the basis for all representations of color. Another very important subject when describing color is
human color vision, which will be discussed briefly in Section 2.3. We then
proceed to an introduction to colorimetry, the study of numerical treatment
of colors, in Section 2.4. Having defined these basics of color, we proceed to issues related to color imaging in Section 2.5, in particular color
management and colorimetric characterization of imaging devices.

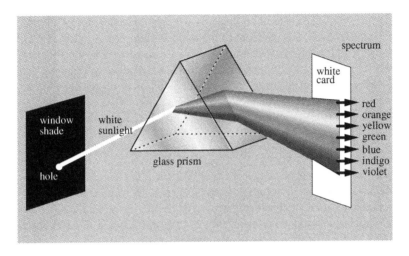

Figure 2.1: *Newton's experiment with sunlight and a prism which led to the realization that the color of light depended on its spectral composition.*

2.2   Light and surfaces

Aristotle viewed all color to be the product of a mixture of white and black, and this was the prevailing belief until Sir Isaac Newton's prism experiments provided the scientific basis for the understanding of color and light (Newton, 1671). Newton showed that a prism could break up white light into a range of colors, which he called the spectrum (see Figure 2.1), and that the recombination of these spectral colors re-created the white light. Although he recognized that the spectrum was continuous, Newton used the seven color names red, orange, yellow, green, blue, indigo, and violet for different parts of the spectrum by analogy with the seven notes of the musical scale. He realized that colors other than those in the spectral sequence do exist, but noted that (Newton, 1730, p.158)

> *All the Colours in the Universe which are made by Light, and depend not on the Power of Imagination, are either the Colours of homogeneal Lights* [i.e., spectral colors], *or compounded of these, . . .*

Light is an important aspect of color. But equally important is the notion of the color of *objects* such as green grass, red roses, yellow submarines, etc. The color of an object is strongly dependent on its spectral reflectance, that is, the amount of the incident light that is reflected from the surface for different wavelengths.[1] If we represent the spectral radiance of the illuminant by the function $l(\lambda)$, λ being the wavelength, and the spectral reflectance in a given surface point of an object by $r(\lambda)$, the radiance of the light reflected from this surface point $f(\lambda)$ is, by definition of reflectance, given in Equation 2.1 and illustrated in Figure 2.2.

$$f(\lambda) = l(\lambda)r(\lambda) \tag{2.1}$$

Note that the model presented in Equation 2.1 is limited in several respects. It does not take into account geometrical effects, for example that the spectral reflectance of an object may depend on the angles of incident light and of observation. One important example of this effect is specular reflection, that is, for a given combination of angle of incidence, surface orientation, and observation angle, the incident light is almost completely reflected, while for other angles, this is not the case. To take into account such effects, the spectral bi-directional reflectance function (SBDRF) should be considered (Nicodemus *et al.*, 1977, Wyszecki and Stiles, 1982, Souami, 1993, Souami and Schmitt, 1995). Additional limitations of this model are its inability to account for effects such as fluorescence, polarization, sub-surface penetration, etc. However, with these limitations in mind, the model of interaction between light and objects presented in Equation 2.1 turns out to be very useful for our further analysis.

2.3 Color vision

In the human eye, an image is formed by light focused onto the retina by the eye's lens. The retina contains two main types of light-sensitive cells, the *rods* and the *cones*. The rods are responsible for night (scotopic) vision and the cones for daylight (photopic) vision under normal levels of illumination. There are three types of cones, named L, M, and S, which are sensitive mainly to light containing long, middle and short wavelengths, respectively (see Figure 2.3). As we will see in Section 2.4, this is the physiological foundation of the so-called *visual trivariance*, which is the

---

[1]Note that a more precise term might be *spectral reflectance factor*. In this document we will not distinguish between the spectral reflectance and the spectral reflectance factor (CIE 17.4, 1989, CIE 15.2, 1986, p.23-24).

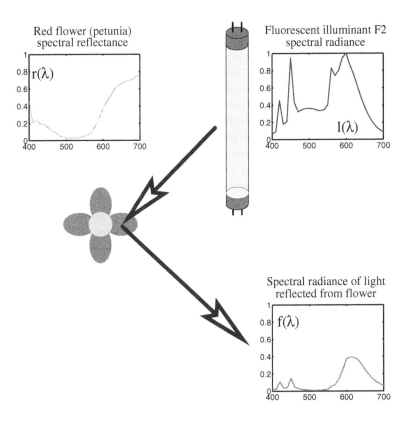

Figure 2.2: *A simple spectral model for the interaction between light and surfaces. The spectral radiance $f(\lambda)$ of the light reflected from a surface with a spectral reflectance $r(\lambda)$, illuminated by an illuminant with spectral radiance $l(\lambda)$ is given by spectral-wise multiplication, $f(\lambda) = l(\lambda)r(\lambda)$.*

basis for our perception of color, and thus also the basis for the colorimetry discussed in this chapter.

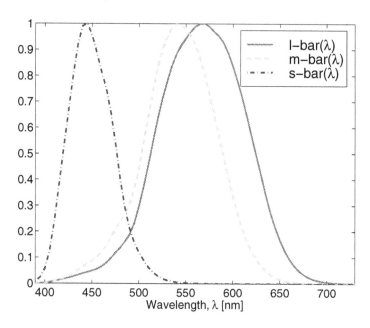

Figure 2.3: *Normalized spectral sensitivity curves $\bar{l}(\lambda)$, $\bar{m}(\lambda)$, and $\bar{s}(\lambda)$, of the three different types of cones, L, M, and S, being responsible for photopic vision, according to Stockman et al. (1993).*

If $f(\lambda)$ is the spectral distribution of light incident on a given location on the retina, the responses of the three cones can be represented as the 3-component vector $\mathbf{c} = [c_1 c_2 c_3]^t$ where

$$c_i = \int_{\lambda_{\min}}^{\lambda_{\max}} f(\lambda) s_i(\lambda) d\lambda, \quad i = 1, 2, 3, \tag{2.2}$$

and $s_i(\lambda)$ denotes the spectral sensitivity of the ith type of cone, and $\lambda_{\min}, \lambda_{\max}$ denote the interval of wavelengths of the visible spectrum outside of which the spectral sensitivities are all zero.

The scope of this dissertation brings us to pay special attention to the color of non-luminous, reflective objects. For such objects, the spectral distribution $f(\lambda)$ of the light incident on the retina is the product of the spectral reflectance $r(\lambda)$ of the object surface and the spectral radiance $l(\lambda)$ of the viewing illuminant, *cf.* Equation 2.1, as shown in Figure 2.4. We may thus rewrite Equation 2.2 as follows.

$$c_i = \int_{\lambda_{\min}}^{\lambda_{\max}} l(\lambda)r(\lambda)s_i(\lambda)d\lambda, \quad i = 1,2,3. \tag{2.3}$$

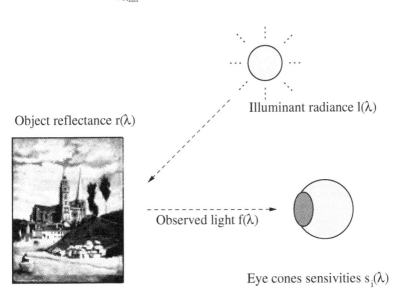

Object reflectance r(λ)

Illuminant radiance l(λ)

Observed light f(λ)

Eye cones sensivities s$_i$(λ)

Figure 2.4: *Human vision of a reflective object. The cone response depends on its spectral sensitivity, the spectral reflectance of the viewed object, and the spectral radiance of the illuminant.*

By uniformly sampling the spectra above with a proper wavelength interval, we can rewrite Equation 2.3 in a matrix form as follows:

$$\mathbf{c} = \mathbf{S}^t \mathbf{L} \mathbf{r}, \tag{2.4}$$

where $\mathbf{S} = [\mathbf{s}_1 \mathbf{s}_2 \mathbf{s}_3]$ is the matrix of eye sensor sensitivities

$$\mathbf{s}_i = [s_i(\lambda_1)s_i(\lambda_2)\ldots s_i(\lambda_N)]^t, \quad \lambda_1 = \lambda_{\min}, \quad \lambda_N = \lambda_{\max},$$

and the sampling interval

$$\delta\lambda = \lambda_i - \lambda_{i-1} = \frac{1}{N-1}(\lambda_{\max} - \lambda_{\min}), \quad i = 2,\ldots,N.$$

$\mathbf{L}$ is the diagonal illuminant matrix with entries from the samples of $l(\lambda)$ along the diagonal, and $\mathbf{r}$ is the sampled spectral reflectance of the object. One of the first to apply such a matrix notation to color issues was Wandell (1987), and this notation has been widely accepted and used since, for example by Jaaskelainen *et al.* (1990), Trussell (1991), Trussell and Kulkarni (1996), Vrhel *et al.* (1994), Sharma and Trussell (1997a).

The spectral sensitivities of the three types of cones define a functional (Hilbert) space, and thus the cone response mechanism corresponds to a projection of the incident spectrum onto the space spanned by the sensitivity functions $s_i(\lambda)$, $i = 1, 2, 3$. This space is called the *Human Visual Sub-Space* (HVSS) (Horn, 1984, Vora and Trussell, 1993). In the sampled case, the HVSS corresponds to the *vector space* spanned by the columns of $\mathbf{S}$.

The cone response functions are quite difficult to measure directly. However, non-singular linear transformations of the cone responses are readily determined through color matching experiments, *cf.* Section 2.4.3. A standardized set of color matching functions $\bar{x}(\lambda)$, $\bar{y}(\lambda)$, and $\bar{z}(\lambda)$ is defined by the CIE (see Section 2.4), and is widely used in colorimetric definitions. The CIE XYZ color matching functions are traced in Figure 2.9 on page 25. Defining $\mathbf{A} = [\bar{x}\bar{y}\bar{z}]$ as the matrix of sampled color matching functions, we can represent a color stimulus using its *CIE XYZ tristimulus values* $\mathbf{t}$ as follows,

$$\mathbf{t} = \mathbf{A}^t \mathbf{f}. \tag{2.5}$$

Note that the linear model of color vision of Equation 2.2 describes only a small part of the complex color-perception process. For example, the model does not explain the intriguing effect of *color constancy*, that is, that the perceived colors of usual objects of vision remain nearly constant independent of the illuminant throughout a wide range, despite the validity of Equation 2.3 (see *e.g.* Hering, 1905, Kroh, 1921, Judd, 1933). In particular, the cone responses cannot be directly related to the common color attributes of *hue*, *saturation* and *lightness*. For a thorough description of the human visual system, refer, for example, to the books of Wandell (1995) or Kaiser and Boynton (1996).

2.4   Colorimetry

In the two previous sections, we have described two aspects of color, the physical aspects of the spectral composition of colored light, and the physiological characteristics of the human visual system. We will now continue to describe the interaction between these two factors, the *psychophysical* aspect of color, as defined by colorimetry. Colorimetry is the branch of color science concerned with the *quantitative specification* of the color of a physically defined visual stimulus in such a manner that (Wyszecki and Stiles, 1982):

1. when viewed by an observer with normal color vision, under the same observing conditions, stimuli with the same specification look alike,

2. stimuli that look alike have the same specification, and

3. the numbers comprising the specification are continuous functions of the physical parameters defining the spectral radiant power distribution of the stimulus.

In colorimetry, we adopt a definition of color that is justified by the experimental fact of visual trivariance, based on the laws of Grassmann which will be described in the following sections.

2.4.1 Grassmann's laws

The laws of Grassmann (1853) represent one of the bases of colorimetry. They are often given in different forms and orders (see *e.g.* Judd and Wyszecki, 1975, Wyszecki and Stiles, 1982, Kowaliski, 1990, Sève, 1996).

Grassmann's first law: Three independent variables are necessary and sufficient to psychophysically characterize a color.

This law states that the color space is three-dimensional. Every color stimulus can be completely matched in terms of three fixed primary stimuli whose radiant powers can be adjusted by the observer to suitable levels. The only restraint on the choice of primaries is that they are *colorimetrically independent*, that is, none of the primaries can be color-matched by a mixture of the two others. The most common set of primaries is red, green and blue. This law is the background for color matching experiments (see Section 2.4.3).

We denote a color stimulus as X, the three primary stimuli as A, B and C, the factors of adjustment proportional to the energy for each primary stimuli as α, β and γ, and a visual equivalence as $\equiv$. We may then express Grassmann's first law as follows:

$$\forall X, \quad \exists \alpha, \beta, \gamma, \quad \text{such that} \quad X \equiv \alpha A + \beta B + \gamma C \qquad (2.6)$$

The three-dimensionality of color is also justified by biological studies of the human eye, as described in Section 2.3.

Grassmann's second law states the principle of color additivity.

Grassmann's second law: The result of an additive mixture of colored light depends only on the psychophysical characterization, and not on the spectral composition of the colors.

The term additive mixture means a color stimulus for which the radiant power in any wavelength interval is equal to the sum of the powers in the same interval of the constituents of the mixture. Using the same notation as before, we may state the law as follows:

$$\forall X_1 \equiv \alpha_1 A + \beta_1 B + \gamma_1 C, X_2 \equiv \alpha_2 A + \beta_2 B + \gamma_2 C,$$
$$X_1 + X_2 \equiv (\alpha_1 + \alpha_2)A + (\beta_1 + \beta_2)B + (\gamma_1 + \gamma_2)C \qquad (2.7)$$

Grassmann's third law may be stated as follows.

Grassmann's third law: If the components of a mixture of color stimuli are moderated with a given factor, the resulting psychophysical color is moderated with the same factor.

That is, if k is a constant,

$$\forall X, k, \quad X \equiv \alpha A + \beta B + \gamma C \Rightarrow kX \equiv k\alpha A + k\beta B + k\gamma C \qquad (2.8)$$

This law implies that all the scales used in colorimetry are continuous.

2.4.2 Tristimulus space

Because of the linear algebraic properties stated by Grassmann's laws, it is possible and convenient to represent color stimuli by vectors in a three-dimensional space, called the *tristimulus space*.

To define this tristimulus space, we need the reference white W, which is defined by the three primaries R, G and B, as $W \equiv \alpha_W R + \beta_W G + \gamma_W B$. We then consider a given color

$$Q \equiv \alpha_Q R + \beta_Q G + \gamma_Q B. \qquad (2.9)$$

Defining the three basis vectors as $\mathbf{r} = \alpha_W R$, $\mathbf{g} = \beta_W G$ and $\mathbf{b} = \gamma_W B$, and denoting the quantities of each of the basis vectors of the primaries as the *tristimulus values* $R_\mathbf{q} = \alpha_Q/\alpha_W$, $G_\mathbf{q} = \beta_Q/\beta_W$ and $B_\mathbf{q} = \gamma_Q/\gamma_W$, the color Q can be defined by the vector $\mathbf{q}$ as follows:

$$\mathbf{q} = R_\mathbf{q}\mathbf{r} + G_\mathbf{q}\mathbf{g} + B_\mathbf{q}\mathbf{b}. \tag{2.10}$$

Once the primary stimuli are defined and fixed, we often represent this equation simply as $\mathbf{q} = [R_\mathbf{q}\, G_\mathbf{q}\, B_\mathbf{q}]^t$. Note that we have now evolved from the term of visual equivalence denoted by $\equiv$ to a simple mathematical equality ($=$).

This vector equation, Eq. 2.10, can be interpreted geometrically, as shown in Figure 2.5 on the next page. The primary stimuli are represented by unit length vectors $\mathbf{r}$, $\mathbf{g}$, and $\mathbf{b}$, with a common origin O. A color stimulus is represented by the tristimulus vector $\mathbf{q}$ whose components have lengths (tristimulus values) $R_\mathbf{q}$, $G_\mathbf{q}$, and $B_\mathbf{q}$ along the directions defined by $\mathbf{r}$, $\mathbf{g}$, and $\mathbf{b}$, respectively. The (r, g, b) trichromatic coordinates are defined by the intersection between the tristimulus vector and the unit plane $(R + G + B = 1)$, giving $r = R_\mathbf{q}/S$, $g = G_\mathbf{q}/S$, and $b = B_\mathbf{q}/S = 1 - (r + g)$, where $S = R_\mathbf{q} + G_\mathbf{q} + B_\mathbf{q}$. The union of a set of colors presented in the two-dimensional representation defined by the equilateral triangle defined by $R + G + B = 1$ is often referred to as the *Maxwell Color Triangle*, see Maxwell (1857) or Wyszecki and Stiles (1982), p. 121. A more convenient representation is the (r, g)-chromaticity diagram in which the r and g coordinate axes are perpendicular to each other.

2.4.3 Color matching

An important notion in colorimetry is *color matching*, referring to visual stimuli typically presented in the two halves of a bipartite visual field, and to judgments of similarities or degrees of difference between the two stimuli. The colorimetric terms are distinguished from the *psychological* terms of color, such as hue, saturation and brightness, which apply to visual concepts that enable the individual observer to describe color perceptions.

Two spectra, represented by the vectors $\mathbf{f}$ and $\mathbf{g}$ produce the same cone reponses, *cf.* Equation 2.4, if

$$\mathbf{S}^t\mathbf{f} = \mathbf{S}^t\mathbf{g}. \tag{2.11}$$

These colors are then said to match. In a color matching experiment (see Figure 2.6), the observer is asked to adjust the amounts of three primary

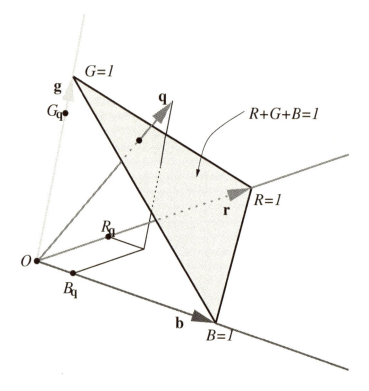

Figure 2.5: *(R,G,B)-tristimulus space. A color stimulus is represented by the tristimulus vector* **q** *whose components have lengths (tristimulus values)* $R_\mathbf{q}$, $G_\mathbf{q}$, *and* $B_\mathbf{q}$ *along the directions defined by the basis vectors* **r**, **g**, *and* **b**, *respectively.*

sources $\mathbf{p}_1$, $\mathbf{p}_2$, and $\mathbf{p}_3$, so that the resulting color matches that of a given light $\mathbf{f}$, that is,

$$\mathbf{S}^t\mathbf{f} = \mathbf{S}^t\mathbf{Pa}, \qquad (2.12)$$

where $\mathbf{P} = [\mathbf{p}_1\mathbf{p}_2\mathbf{p}_3]$ denotes the primaries and $\mathbf{a} = [a_1 a_2 a_3]^t$ corresponds to the three weights. It can be shown that if the primaries are colorimetrically independent, the vector of weights exists and is equal to $\mathbf{a} = (\mathbf{S}^t\mathbf{P})^{-1}\mathbf{S}^t\mathbf{f}$. However, for a given spectrum $\mathbf{f}$, the vector of weights may take values. Since negative intensities of the primaries cannot be produced, the spectrum $\mathbf{Pa}$ is not realizable using the primaries. In practice

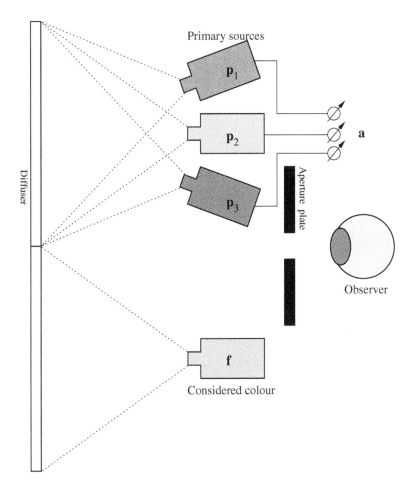

Figure 2.6: *Principle of trichromatic color matching by additive mixing of lights. The observer views a small circular field which is split into two halves, one on which the color* **f** *which is to be matched is displayed, the other displaying an additive mixture of the three primary sources* **p**$_1$, **p**$_2$, *and* **p**$_3$, *typically red, green and blue.*

a color matching experiment is arranged by mixing the primaries having negative strengths with the considered color, instead of with the other primaries. This might be represented, for example, as matching $a_1\mathbf{p}_1 + a_2\mathbf{p}_2$ with $(-a_3)\mathbf{p}_3 + \mathbf{f}$ in the case where a_3 is negative.

2.4.4 Color matching functions

If color matching experiments are conducted with the set of stimuli $\mathbf{e}_i$, $i = 1, \ldots, N$ being monochromatic light of varying wavelengths and constant unit energy, we may obtain the weights $\mathbf{a}_i$ for each wavelength. Doing this for all the N wavelengths of the sampling interval that is used, we may combine the color matching results into one equation,

$$\mathbf{S}^t\mathbf{I} = \mathbf{S}^t\mathbf{P}\mathbf{A}^t, \tag{2.13}$$

where $\mathbf{I} = [\mathbf{e}_1\mathbf{e}_2 \ldots \mathbf{e}_N]$ is the $(N \times N)$ unit matrix and $\mathbf{A} = [\mathbf{a}_1\mathbf{a}_2 \ldots \mathbf{a}_N]^t$ is the *color matching matrix* corresponding to the primaries $\mathbf{P}$. The columns of $\mathbf{A}$ are referred to as the *color matching functions* associated with the primaries $\mathbf{P}$. Since any spectrum can be represented as a linear combination of the unit spectra, its color tristimulus values can be readily calculated as $\mathbf{t} = \mathbf{A}^t\mathbf{f}$, *cf.* Equation 2.5.

2.4.5 Metamerism

From Equation 2.11, $\mathbf{S}^t\mathbf{f} = \mathbf{S}^t\mathbf{g}$, and the fact that $\mathbf{S}$ is a $N \times 3$ matrix, $N > 3$, it is clear that there are several different spectra that can appear as the same color to the observer. A set of two such spectra having different spectral compositions but giving rise to the same psychophysical characterization are called *metamers* (CIE 15.2, 1986, CIE 80, 1989). An example of metamerism is given in Figure 2.7.

Metamerism implies that two objects which appear to have exactly the same color, may have very different colors under different lighting conditions. The color mismatch due to loss of metamerism when changing observer or lighting can be predicted numerically (see *e.g.* Ohta and Wyszecki, 1975, Schmitt, 1976, Wyszecki and Stiles, 1982, ch. 3.8.5). This may be an important practical problem, *e.g.* in the clothing industry, where the colors of fabrics of different types should match, both inside and outside the store.

Notice, however, that even though it may cause some problems, metamerism is the basis of the entire science of color. Without metamerism,

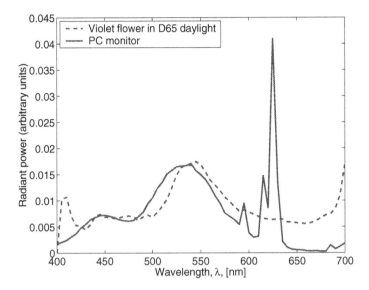

Figure 2.7: *Spectral radiant power distributions of daylight (D65) reflected from a violet flower and emitted by a computer monitor tuned to match the color of the flower. The two spectra are metamers.*

there would be no colorimetry and no color image reproduction on paper or screen as we know it. The only possible way of reproducing images would be to recreate the spectral reflectance of the original objects, creating a *spectral match* as opposed to a *metameric match*.

2.4.6 CIE standard illuminants

The most important of all color specification systems is that developed by the Commission Internationale de l'Eclairage (CIE). It provides a standard method for describing the stimulus of a color, under controlled lighting and viewing conditions, based on the average known response of the human visual system. It is derived from careful psychophysical experiments and is thoroughly documented. The CIE system has the force of an international standard, and has become the basis of all industrial colorimetry.

Because the appearance of a color strongly depends on the color of the illuminant, it is clear that an essential step in specifying color is an accurate definition of the illuminants involved. In 1931, the CIE recommended the use of three standard illuminants, denoted A, B and C, whose spectral power distribution curves are shown in the left part of Figure 2.8. Standard illuminant A consists of a tungsten filament lamp at a given color temperature, while B and C consist of A together with certain liquid color filters (Wyszecki and Stiles, 1982, p. 148). A is intended to be representative of tungsten filament lighting, B of direct sunlight, and C of light from an overcast sky.

However, even if the illuminants B and C fairly well represent the spectral power distribution of daylight over most of the spectrum, they are seriously deficient at wavelengths below 400 nm. Due to the increasing use of dyes and pigments that have fluorescent properties, the CIE later defined several power distributions representing daylight at all wavelengths between 300 and 830 nm. In the right part of Figure 2.8 on the facing page the distributions D_{50} and D_{65} are shown. D_{65} represents a standard daylight for general use, and D_{50} is somewhat more yellow. The subscripts 50 and 65 refer to the color temperature of the illuminants, *e.g.* D_{50} has a correlated color temperature[2] of 5000K. In addition to these sources, the hypothetical equi-energetic illuminant E, having equal radiance power per unit wavelength throughout the visible spectrum, is also defined.

These standard illuminants are widely used in color systems and standards. In television, D_{65} is the reference white for PAL and C for the NTSC television system. D_{50} is extensively used in the graphic arts industry.

2.4.7 CIE standard observers

There are slight differences in the amounts of color stimuli required to obtain a given color perception between different observers. Some of these differences are random, and disappear if the results of several tests by each observer are averaged. But there remain some discrepancies that must be attributed to differences in the color vision of the individual observers.

In 1931 the CIE defined a *standard observer*, based on experimental results obtained by W. D. Wright and J. Guild, and by K. S. Gibson and E. P. T. Tyndall, see *e.g.* Chapter 8 of Hunt (1995) or Chapter 3 of Wyszecki and Stiles (1982).

---

[2]The correlated color temperature is defined as the temperature of the Planckian radiator whose perceived color most closely matches that of a given stimulus seen at the same brightness and under specified viewing conditions (Hunt, 1991, CIE 17.4, 1989).

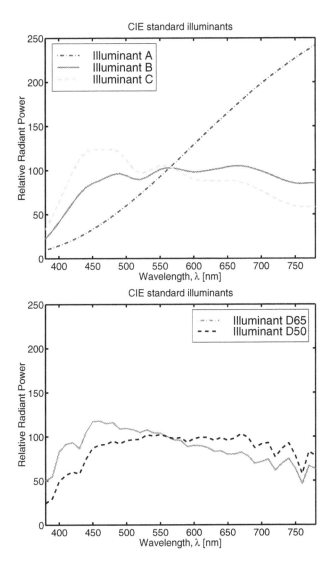

Figure 2.8: *Relative spectral power distributions of the standard illumi-nants A, B, and C (above) and D_{65} and D_{50} (below).*

These standard-observer data consist of the color matching functions obtained with the monochromatic primaries of wavelengths $R_0 = 700$ nm, $G_0 = 546.1$ nm, and $B_0 = 435.8$ nm, and for the reference equi-energetic white E. The color matching functions for the standard observer are sketched in the left part of Figure 2.9 on the next page. From these functions, given the spectral power distribution curve of any color, it is possible to calculate the amount of the three stimuli required by the standard observer to match a given color, *cf.* Section 2.4.3. This defines the CIE 1931 Standard RGB Colorimetric System, which is a basis in colorimetry. A given color stimulus with spectral radiant power distribution $f(\lambda)$ can be represented as three RGB tristimulus values by the following formulae:

$$R = \int_{\lambda_{\min}}^{\lambda_{\max}} f(\lambda)\bar{r}(\lambda)d\lambda \tag{2.14}$$

$$G = \int_{\lambda_{\min}}^{\lambda_{\max}} f(\lambda)\bar{g}(\lambda)d\lambda \tag{2.15}$$

$$B = \int_{\lambda_{\min}}^{\lambda_{\max}} f(\lambda)\bar{b}(\lambda)d\lambda. \tag{2.16}$$

The CIE 1931 Standard XYZ Colorimetric System is defined in a similar manner, using the color matching functions $\bar{x}(\lambda)$, $\bar{y}(\lambda)$ and $\bar{z}(\lambda)$, shown in Figure 2.9. The tristimulus values X, Y and Z are defined as follows:

$$X = \int_{\lambda_{\min}}^{\lambda_{\max}} f(\lambda)\bar{x}(\lambda)d\lambda \tag{2.17}$$

$$Y = \int_{\lambda_{\min}}^{\lambda_{\max}} f(\lambda)\bar{y}(\lambda)d\lambda \tag{2.18}$$

$$Z = \int_{\lambda_{\min}}^{\lambda_{\max}} f(\lambda)\bar{z}(\lambda)d\lambda. \tag{2.19}$$

The set of color matching functions $\bar{x}(\lambda)$, $\bar{y}(\lambda)$ and $\bar{z}(\lambda)$ is a linear transformation of the set $\bar{r}(\lambda)$, $\bar{g}(\lambda)$ and $\bar{b}(\lambda)$, as follows:

$$\begin{bmatrix} \bar{x}(\lambda) \\ \bar{y}(\lambda) \\ \bar{z}(\lambda) \end{bmatrix} = \begin{bmatrix} 0.49 & 0.31 & 0.2 \\ 0.17697 & 0.81240 & 0.01063 \\ 0.0 & 0.01 & 0.99 \end{bmatrix} \cdot \begin{bmatrix} \bar{r}(\lambda) \\ \bar{g}(\lambda) \\ \bar{b}(\lambda) \end{bmatrix} \tag{2.20}$$

Note that the XYZ color matching functions do not correspond to a set of physical primaries, as was the case with the RGB color matching functions defined above. They correspond to three non-physical primaries with

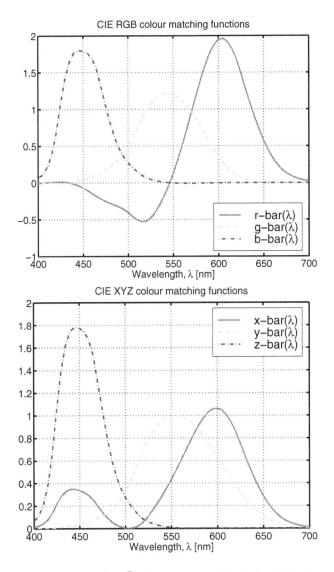

Figure 2.9: *CIE* $\bar{r}(\lambda)$, $\bar{g}(\lambda)$, $\bar{b}(\lambda)$ *(left) and* $\bar{x}(\lambda)$, $\bar{y}(\lambda)$, $\bar{z}(\lambda)$ *(right) color matching functions.*

the reference equi-energetic white E, chosen so that the color matching functions have only positive values. The numbers in the matrix of Equation 2.20 were carefully chosen by the CIE to ensure that the tristimulus values X, Y and Z are all positive and so that the value of Y is proportional to the luminance of the given color.

To graphically visualize a color, the CIE (x, y) chromaticity diagram (see Figure 2.10) is often used. The x and y values are tristimulus values normalized such that $x + y + z = 1$, *cf.* Section 2.4.2.

2.4.8 Uniform color spaces and color differences

Psychophysical experiments have shown that the human eye's sensitivity to light is not linear. The RGB and XYZ color spaces defined by the CIE are related linearly to the spectral power distribution of the colored light.

When changing the tristimulus values XYZ (or RGB) of a color stimulus, the observer will perceive a difference in color only after a certain amount, equal to the Just Noticeable Difference (JND). In both RGB and XYZ spaces the JND depends on the location in the color space.

These are two major drawbacks of the color spaces presented in the previous section. To remedy this, the CIE proposed in 1976 two *pseudo-uniform*[3] color spaces, denoted CIELUV and CIELAB (see CIE 15.2, 1986). The CIELUV space was often used for describing colors in displays, while CIELAB was initially designed for reflective media. Now CIELAB is used for most applications and has been chosen as standard color space for several fields, *e.g.* in graphic arts (ISO 12639, 1997), multimedia (IEC 61966-8, 1999), color facsimile (ITU-T T.42, 1994). In our work, we make extensive use of the CIELAB space, and we will therefore describe it in detail in the following section.

---

[3] A color space is called *uniform*, or *psychometric*, when equal visually perceptible differences are produced with equi-spaced points throughout the space, that is, the JND is constant throughout the entire color space. In practice, this condition is only fulfilled approximately, thus we normally use the term *pseudo-uniform*. Notice that the notion of JND is observer-dependent and somewhat subjective. CIE's color spaces are based on a standard observer.

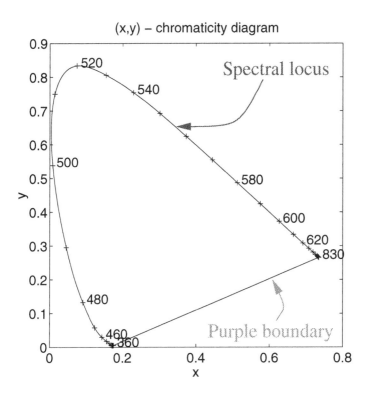

Figure 2.10: *Chromaticity diagram* (x, y) *of CIE 1931 XYZ standard colorimetric observer. The curved line shows where the colors of the spectrum lie and is called the spectral locus; the wavelengths are indicated in nanometers along the curve. If two colors are additively mixed together, then the point representing the mixture is located in the diagram by a point that always lies on the line joining the two points representing the original colors. This means that, if the two ends are joined by a straight line, that line represents mixtures of light from the two ends of the spectrum; as those colors are mixtures of red and blue, this line is known as the purple boundary. The area enclosed by the spectral locus and the purple boundary encloses the domain of all possible colors.*

2.4.8.1 CIELAB color space

The CIELAB pseudo-uniform color space is defined by the quantities L^*, a^* and b^*, defined as follows:[4]

$$L^* = 116 f(\frac{Y}{Y_n}) - 16 \qquad (2.21)$$

$$a^* = 500 \left[f(\frac{X}{X_n}) - f(\frac{Y}{Y_n}) \right] \qquad (2.22)$$

$$b^* = 200 \left[f(\frac{Y}{Y_n}) - f(\frac{Z}{Z_n}) \right] \qquad (2.23)$$

where

$$f(\alpha) = \begin{cases} \alpha^{\frac{1}{3}} & , \alpha \geq 0.008856 \\ 7.787\alpha + \frac{16}{116} & , \text{otherwise} \end{cases}$$

The tristimulus values X_n, Y_n and Z_n are those of the nominally white stimulus. For the example of illuminant D_{50} the values are calculated as follows:

$$Y_n = \int_{\lambda_{\min}}^{\lambda_{\max}} 1 \cdot l_{D_{50}}(\lambda) \cdot \bar{y}(\lambda) d\lambda = 96.42 \qquad (2.24)$$

$$X_n = \int_{\lambda_{\min}}^{\lambda_{\max}} 1 \cdot l_{D_{50}}(\lambda) \cdot \bar{x}(\lambda) d\lambda = 100.00 \qquad (2.25)$$

$$Z_n = \int_{\lambda_{\min}}^{\lambda_{\max}} 1 \cdot l_{D_{50}}(\lambda) \cdot \bar{z}(\lambda) d\lambda = 82.49 \qquad (2.26)$$

L^* represents the *lightness* of a color, known as the CIE 1976 psychometric lightness. The scale of L^* is 0 to 100, 0 being the ideal black, and 100 being the reference white. The chromaticity of a color can be represented in a two-dimensional (a^*, b^*) diagram (see Figure 2.11(b)), a^* representing the degree of green versus red, and b^* the degree of blue versus yellow. Note that, in contrast to the (x, y) chromaticity diagram (Fig. 2.10), a mixture of two colors is not necessarily situated on the straight line joining the two colors. The (a^*, b^*) chroma diagram is *not* a chromaticity diagram.

---

[4]The asterisks are used mostly for historical reasons, and we will sometimes omit them to simplify notation.

An alternative representation of colors in the CIELAB space emanates when using cylindrical coordinates, defining the *CIE 1976 chroma*, designating the distance from the L^*-axis, as

$$C_{ab}^* = \sqrt{a^{*2} + b^{*2}}, \qquad (2.27)$$

and the *CIE 1976 hue-angle*,

$$h_{ab} = \arctan\left(\frac{b^*}{a^*}\right). \qquad (2.28)$$

The use of these quantities, lightness L^*, chroma C_{ab}^*, and hue angle h_{ab} may facilitate the intuitive comprehension of the CIELAB color space, by relating it to perceptual attributes of colors.

An illustration of the uniformity of the CIELAB color space is shown in Figure 2.11, where we compare the loci of constant hue and chroma according to Munsell in the xy and the a^*b^* planes. We see that the loci are far more distorted in the CIE 1931 (x, y) chromaticity diagram than in the (a^*, b^*) chroma diagram. We note, however, that the CIELAB space is not perfectly uniform.

2.4.8.2   Color difference formulae

When comparing two colors, specified by $[L_1^*, a_1^*, b_1^*]$ and $[L_2^*, a_2^*, b_2^*]$, one widely used measure of the color difference is the *CIE 1976 CIELAB color-difference* which is simply calculated as the Euclidean distance in CIELAB space, as follows:

$$\Delta E_{ab}^* = \sqrt{(L_1^* - L_2^*)^2 + (a_1^* - a_2^*)^2 + (b_1^* - b_2^*)^2} \qquad (2.29)$$

The interpretation of ΔE_{ab}^* color differences is not straightforward, though. It is commonly stated (Kang, 1997) that the JND is equal to 1. However Mahy *et al.* (1994a) found a JND of $\Delta E_{ab}^* = 2.3$. A rule of thumb for the practical interpretation of ΔE_{ab}^* when two colors are shown side by side is presented in Table 2.1 on page 31. Another interpretation of ΔE_{ab}^* errors for the evaluation of scanners is proposed by Abrardo *et al.* (1996). They classify mean errors of 0-1 as *limit of perception*, 1-3 as *very good quality*, 3-6 as *good quality*, 6-10 as *sufficient*, and more than 10 as *insufficient*. We note the disagreement between these classifications, underlining the fact that the evaluation of quality and acceptability is highly subjective and depends on the application.

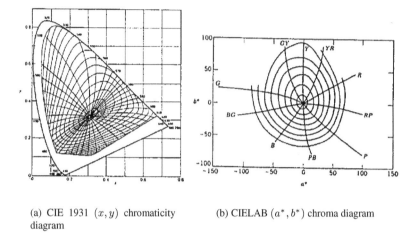

(a) CIE 1931 (x, y) chromaticity diagram

(b) CIELAB (a^*, b^*) chroma diagram

Figure 2.11: *Munsell loci of constant hue and chroma (from Billmeyer and Saltzman, 1981). We see that the loci are far more distorted in the CIE 1931 (x, y) chromaticity diagram (a) than in the (a^*, b^*) chroma diagram (b). This illustrates the fact that the CIELAB color space is more perceptually uniform than the XYZ color space.*

It may also be interesting to evaluate the differences of each of the components of the CIELAB space separately. This is straightforward for L^*, a^*, b^*, and C_{ab}^*, however, for the hue angle h_{ab} this merits some special consideration. Of course, the direct angle difference in degrees may be instructive. However, to allow color differences to be broken up into components of lightness, chroma and hue, whose squares sum to the square of ΔE_{ab}^*, a quantity ΔH^*, called the *CIE 1976 hue-difference*, is defined as

$$\Delta H^* = \sqrt{(\Delta E_{ab}^*)^2 - (\Delta L^*)^2 - (\Delta C_{ab}^*)^2}. \qquad (2.30)$$

The color difference formula of Equation 2.29 is supposed to give a measure of color differences that is perceptually consistent. However, since it has been found that the CIELAB space is *not* completely uniform, the color difference ΔE_{ab}^* is not perfect. Several attempts have been made to define better color difference formulae, *e.g.* the CMC formula (Clarke *et al.*, 1984, McLaren, 1986) and the BFD formula (Luo and Rigg,

Table 2.1: *Rule of thumb for the practical interpretation of* ΔE_{ab}^* *measuring the color difference between two color patches viewed side by side.*

ΔE_{ab}^*	Effect
< 3	Hardly perceptible
$3 < 6$	Perceptible, but acceptable
> 6	Not acceptable

1987a;b). A comparison of these and other uniform color spaces using perceptibility and acceptability criteria is done by Mahy *et al.* (1994a).

Recently, the CIE defined the *CIE 1994 color-difference model* (McDonald and Smith, 1995), abbreviated CIE94, denoted ΔE_{94}^*, based on the CIELAB space and the previously cited works on color difference evaluation. They defined reference conditions under which the new metric, with default parameters, is expected to perform well:

1. The specimens are homogeneous in color.

2. The color difference ΔE_{ab}^* is less than 5 units.

3. They are placed in direct edge contact.

4. Each specimen subtends an angle of more than 4 degrees to the assessor, whose color vision is normal.

5. They are illuminated at 1000 lux, and viewed against a background of uniform gray, with $L^* = 50$, under illumination simulating D65.

The color difference is calculated as a weighted mean-square sum of the differences in lightness, ΔL^*, chroma, ΔC^*, and hue, ΔH^*.

$$\Delta E_{94}^* = \sqrt{\left(\frac{\Delta L^*}{k_L S_L}\right)^2 + \left(\frac{\Delta C^*}{k_C S_C}\right)^2 + \left(\frac{\Delta H^*}{k_H S_H}\right)^2} \qquad (2.31)$$

The *weighting functions* S_L, S_C, and S_H vary with the chroma of the reference specimen[5] C^* as follows,

$$S_L = 1, \quad S_C = 1 + 0.045 C^*, \quad S_H = 1 + 0.015 C^*. \qquad (2.32)$$

[5] If neither of the two samples can be considered to be a reference specimen, the geometric mean of the chroma of the two samples is used.

The variables k_L, k_C and k_H are called *parametric factors* and are included in the formula to allow for adjustments to be made independently to each color difference term to account for any deviations from the reference viewing conditions, that cause component specific variations in the visual tolerances. Under the reference conditions explained above, they are set to

$$k_L = k_C = k_H = 1. \tag{2.33}$$

We note that under reference conditions, ΔE_{94}^* equals ΔE_{ab}^* for neutral colors, while for more saturated colors, ΔE_{94}^* becomes smaller than ΔE_{ab}^*.

This color difference formula is now extensively used both in literature and industry, and is expected to replace ΔE_{ab}^* as the most popular way of expressing color differences.

2.5 Color imaging

The main subject of this dissertation is color imaging. Especially important is color consistency throughout a color imaging system. To achieve this, it is necessary to understand and control the way in which the different devices involved in the entire color imaging chain treat colors. We will mainly be concerned with *digital* imaging, in which the different devices are connected to a computer, as illustrated in Figure 2.12. Our goal is to make sure that all these devices work properly together.

We will first present the concept of *color management*, providing a framework in which color information can be processed consistently throughout a digital imaging system. Then we proceed to a brief presentation of digital image acquisition and reproduction devices. It is not in the scope of this thesis to describe in detail the different technologies used in such devices. We will, however, concentrate on how they can be characterized colorimetrically.

2.5.1 Color management

Whenever a computer is used for the acquisition, visualization, or reproduction of colored objects, it is important to assure color consistency throughout the system (Hardeberg and Schmitt, 1998). By calibrating color peripherals to a common standard, Color Management System (CMS) software makes it easier to match the colors that are scanned to those that appear on the monitor and printer, and also to match colors designed on

Figure 2.12: *Different digital imaging devices connected to a central computer. A typical imaging workflow goes from an original document, scanned, visualized on the monitor, and finally printed. Ideally, the printed result should be an exact facsimile of the original document.*

the monitor, using for example CAD software, to the printed document. Color management is highly relevant to persons using computers for working with art, architecture, desktop publishing or photography, but also to non-professionals, as for example, when displaying and printing images downloaded from the Internet or from a Photo CD (Photo CD, 1991).

But where is the problem in all this? For example, one might say: "I know that my scanner provides me with a description of each color as a unique combination of red, green, and blue (RGB) and so does my monitor, and even my ink-jet printer accepts RGB images!" The problem is that even if these devices all 'speak' RGB, the way they describe colors (scanner-RGB, monitor-RGB and printer-RGB) are substantially different, even for peripherals of the same type. An obvious example of this is that an image printed on glossy paper by a sublimation printer is considerably more colorful than the same image printed on plain paper by an old ink-jet printer.

To obtain faithful color reproduction, a Color Management System (CMS) has two main tasks. First, colorimetric characterization of the peripherals is needed, so that the *device-dependent* color representations of the scanner, the printer, and the monitor can be linked to a *device-independent* color space, the Profile Connection Space (PCS), see Figure 2.13. This is the process of *profiling*. Furthermore, efficient means for processing and converting images between different representations are needed. This task is undertaken by the Color Management Module (CMM), see Figure 2.14. For further information about the architecture of CMS, refer *e.g.* to MacDonald (1993a), Murch (1993), Schläpfer *et al.* (1998), ICC.1:1998.9 (1998). In Figure 2.15 we present an example of a color management system for a color facsimile system.

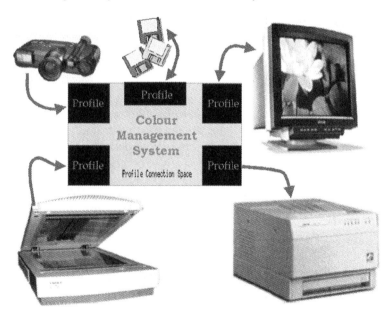

Figure 2.13: *Different digital imaging devices connected in a color management system. Each device is characterized by a profile. Note the workflow simplification compared to Figure 2.12 on the preceding page.*

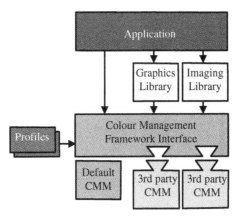

Figure 2.14: *A typical Color Management System architecture, as described in the ICC Profile Format Specification (ICC.1:1998.9, 1998).*

The industry adoption of CMS depends strongly on standardizations (Stokes, 1997). The International Color Consortium [6] (ICC) plays a very important role in this concern. The ICC was established in 1993 by eight industry vendors for the purpose of creating, promoting and encouraging the standardization and evolution of an open, vendor-neutral, cross-platform color management system architecture and components. Today there is wide acceptance of the ICC standards.

Several vendors offer CMS software solutions, for example the following:[7]

- Agfa Gevaert N.V. with ColorTune,
  http://www.agfa.com/software/colortune.html

- Apple Inc. with ColorSync,
  http://www.apple.com/colorsync/

- CCE S.A.R.L. with AffixColor
  http://www.affixcce.com/

- Eastman Kodak Company with ColorFlow,
  http://www.kodak.com/go/colorflow

---

[6] See http://www.color.org for more information about the ICC.

[7] See http://color.hardeberg.com/cms.html for a more comprehensive list of available Color Management System software.

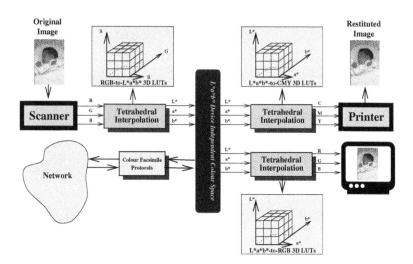

Figure 2.15: *An example of a color management system for color fac-simile (Hardeberg et al., 1996). The transformations between the device-dependent color coordinates (RGB and CMY) and the CIELAB color space are performed using 3D look-up tables and a tetrahedral interpolation technique (cf. Appendix B). The look-up tables are determined by the characterization algorithms.*

- E-Color, Inc. with Colorific,
  http://www.ecolor.com

- FotoWare AS with Color Factory,
  http://www.fotoware.com

- Heidelberg CPS GmbH with LinoColor,
  http://www.linocolor.com

- Imaging Technologies Corporation with ColorBlind,
  http://www.color.com

- GretagMacbeth with ProfileMaker,
  http://www.gretagmacbeth.com

It has been concluded in a recent study (Schläpfer *et al.*, 1998) under-taken by the Association for the Promotion of Research in the Graphic

Arts Industry (UGRA) that the color management solutions offered by different vendors are approximately equal, and that color management now has passed the breakthrough phase and can be considered a valid and useful tool in image reproduction.

However, there is still a long way to go, both when it comes to software development (integration of CMS in operating systems, user-friendliness, simplicity, ...), research in color and imaging science and technology (better color consistency, gamut mapping, color appearance models, ...), and standardization. Color imaging is a very active research domain, and in the next sections, we will briefly review different approaches to the colorimetric characterization of image acquisition and reproduction devices.

2.5.2 Digital image acquisition

In order to process images digitally, the continuous-space, analog, real-world images need to be sampled and quantized. This is typically done by a digital camera or scanner. There have been significant improvement in the quality of digital image acquisition devices over the last several years, and at the same time, prices are reduced dramatically. Traditional analog imaging is constantly loosing market shares. However, there are several technical issues that still need to be solved in digital image acquisition. In Chapter 4 we present our approach to the acquisition of high quality digital color images. A very important problem is how to attain a high colorimetric fidelity, and this issue is addressed in Chapter 3.

2.5.2.1 Colorimetric characterization of scanners and cameras

To colorimetrically characterize image acquisition devices such as CCD cameras and scanners, two different approaches are typically used, applying spectral and analytical models. The goal of a *spectral* characterization technique will be to estimate the function $f(\cdot)$ in Equation 2.34, this function representing the transformation performed by the scanner, from object reflectance $r(\lambda)$ to scanner RGB values, that is, the spectral model of the scanner. Eventually, this information can be used to obtain device-independent color information by defining an "optimal" function $g(\cdot)$. If $f(\cdot)$ meets the *Luther-Ives condition* (Ives, 1915, Luther, 1927) this is trivial. Otherwise we have to define what is meant by "optimal".

A scanner characterization based on *analytical* models, however, seeks to minimize the difference between the known device-independent CIE-

LAB values $(\text{Lab})_t$ of the color patches of a target and the values $(\text{Lab})_c$ as obtained by the desired transformation $g(\cdot)$ from the scanner RGB values. Notice that a device-independent color representation other than CIELAB may be used.

$$
\begin{array}{ccc}
 & f(\cdot) & \\
r(\lambda) & \to & \text{RGB} \\
h(\cdot) \quad \downarrow & & \downarrow \quad g(\cdot) \\
(\text{Lab})_t & \overset{?}{=} & (\text{Lab})_c \\
 & \Delta E &
\end{array}
\tag{2.34}
$$

Analytical models. For the colorimetric characterization of electronic image input devices, it is current practice to use standard color targets such as the ANSI IT8.7/2 (1993) chart and to apply analytical models for the mapping of the input device data into a standardized device-independent color space. The mapping function is typically obtained by polynomial regression, see *e.g.* Berns (1993a), Lenz *et al.* (1996b), Hardeberg *et al.* (1996), as well as the surveys by Johnson (1996) and Kang (1997). Quite often, the transformation from scanner RGB to CIEXYZ is performed using a 3×3 matrix.

An important limitation of such methods is that, for a given experimental setup of the lighting conditions and for a given choice of the illuminant, individual characterization data have to be obtained for each type of input media, the failure to do this resulting in considerable errors due to metamerism. However, for a given input medium, such methods give very satisfactory results. We report on our approach to the analytical colorimetric characterization of desktop scanners in Chapter 3.

Spectral models For a more complete characterization, the knowledge of the physical characteristics of the different optical and electro-optical components that are involved in the image conversion process would be desirable. This is particularly the case for applications where the camera will be used for the acquisition of multispectral images. A simple spectral model of the image acquisition process may be formulated in terms of algebraic matrix operations. The spectral characterization consists in estimating the different spectral characteristics of the sensor, the optics and the illumination, or eventually, the joint characteristics of these elements (see *e.g.* Farrell and Wandell, 1993, Sharma and Trussell, 1996c, Hardeberg *et al.*, 1998b).

In Section 6.2, we investigate an approach to this problem based on the acquisition of a number of samples with known reflectance spectra. By observing the camera output to known input, we perform an estimation of the spectral sensitivity of a CCD camera.

2.5.3 Digital image reproduction

Color may be produced in many different ways. According to Nassau (1983), as many as fifteen distinct physical mechanisms are responsible for color in nature. Only few of these mechanisms are suitable for digital image reproduction, but there exists nevertheless considerable diversity in available technologies for displaying and printing color images. Image reproduction devices can be broadly classified into two categories, *additive* and *subtractive* devices. Some devices also combine these two technologies; they are called *hybrid* devices.

2.5.3.1 Additive color devices

In additive color devices, the colors are produced by adding light of different colors, following the theories of additive color mixture described earlier in this chapter. The most common choice of additive primary colors is red, green and blue (RGB).

Visual display units[8] (VDU) emit light and are therefore additive devices. They can be characterized almost completely in terms of a few parameters, such as the white point, the gamma curve etc. When these parameters are known, the required RGB drive signals needed to produce a given XYZ color stimulus can be calculated, see *e.g.* NPL QM 117 (1995), Berns *et al.* (1993), and Chapter 14 of Kang (1997).

Recently, a new standard color space was proposed, the sRGB color space[9] (Anderson *et al.*, 1996, IEC 61966-2.1, 1999). Its definition is based on the average performance of PC displays under normal viewing conditions. We present here the steps involved in the conversions between CIEXYZ and sRGB as an example. If exact colorimetric reproduction is needed on a particular VDU, formulas resembling the following should be

---

[8]Visual display units, display, monitor, and computer screen, are different names used for this device. Two important types are Cathode-Ray Tubes (CRT) and Liquid Crystal Displays (LCD).

[9]See http://www.srgb.com for more information on how the use of the sRGB color space can facilitate color consistency, as a simpler alternative to ICC-based color management.

used, but with different parameters, obtained from a colorimetric characterization of the device.

The sRGB tristimulus values are defined simply as a linear transformation of the CIEXYZ values, based on phosphor chromaticities and D_{65} white point, as follows,

$$
\begin{bmatrix} R_{\mathrm{sRGB}} \\ G_{\mathrm{sRGB}} \\ B_{\mathrm{sRGB}} \end{bmatrix} = \begin{bmatrix} 3.2406 & -1.5372 & -0.4986 \\ -0.9689 & 1.8758 & 0.0415 \\ 1.0570 & -0.2040 & 0.0557 \end{bmatrix} \begin{bmatrix} X \\ Y \\ Z \end{bmatrix}. \quad (2.35)
$$

Then the non-linear sR'G'B' values are defined as

$$
R'_{\mathrm{sRGB}} = \begin{cases} 12.92 R_{\mathrm{sRGB}}, & R_{\mathrm{sRGB}} \leq 0.00304 \\ 1.055 R_{\mathrm{sRGB}}^{1.0/2.4} - 0.055, & \text{elsewhere}, \end{cases} \quad (2.36)
$$

and likewise for G'_{sRGB} and B'_{sRGB}. The 8-bit digital values that should be transmitted to the display are finally calculated as

$$
R_{\mathrm{8bit}} = 255.0 R'_{\mathrm{sRGB}}.
$$

2.5.3.2 Subtractive color devices

In additive color devices, the colors are typically produced by adding different proportions of the three primary colors red, green and blue. In subtractive color devices, the colors are produced by multiplying a white spectrum by the spectral transmission curves $\tau(\lambda)$ of the three subtractive primary colors cyan, magenta and yellow (CMY). Thus, for each of the subtractive primaries, frequency components are removed from the white spectrum. An ideal subtractive color device can be presented as in Figure 2.16, where we observe that the result of a multiplication of an ideal white spectrum with the three ideal rectangular bandstop filters gives a resulting color spectrum exactly equal to the one obtained in an ideal additive system.

We remark that no concepts in the field of color have traditionally been more confused than that of additive and subtractive color mixture. This confusion can be traced to two prevalent misnomers: the subtractive primary cyan, which is properly a blue-green, is commonly called blue; and the subtractive primary magenta is commonly called red. In these terms, the subtractive primaries become red, yellow, and blue; and those whose experience is confined for the most part to subtractive mixtures have good

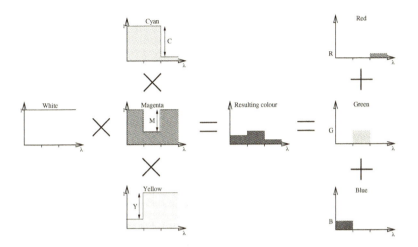

Figure 2.16: *An ideal subtractive color reproduction system. We see that the resulting color of a subtractive color system results from a multiplication of a white spectrum with the spectra of yellow, magenta and cyan inks. We note that the resulting spectrum equals the sum of three ideal additive RGB primaries which are the spectral complimentaries of the ideal inks.*

cause to wonder why the physicist insists on regarding red, *green*, and blue as the primary colors. The confusion is at once resolved when it is realized that red, green, and blue are selected as additive primaries because they provide the greatest color gamut in mixtures. For the same reason, the subtractive primaries are, respectively, red-absorbing (cyan), green-absorbing (magenta), and blue-absorbing (yellow).

The principle of subtractive color mixture is used in color printers, where a white sheet is covered with layers of yellow, magenta and cyan inks or other materials. The pigments in the inks absorb certain wavelengths from the incident light, and thus constitute a subtractive color system.

The input to a typical printer is a quadruple $[C, M, Y, K]$. The C, M and Y represents the amount of cyan, magenta and yellow ink, while K represents the black ink, often denoted as the black separation. The black separation is introduced to accomplish two things: to increase the contrast by increasing the density in the dark areas of the picture, and to replace some percentage of the three primaries for economic or mechanical reasons, as explained *e.g.* by Stone *et al.* (1988). There are several strategies for how the amount of black ink is determined. One is gray-component replacement (GCR), in which the neutral or gray component of a three-color image is replaced with a certain level of black ink. The least predominant of the three primary inks is used to calculate a partial or total substitution by black, and the color components of the image are reduced to produce a print image of a nearly equivalent color to the original three-color print (Sayanagi, 1986, Johnson, 1992, Kang, 1997). In the further, we will often omit the K when describing the output to a printer, as we treat this black separation as a device characteristic.

The relation between the ideal components C, M, Y and the RGB-space based on ideal bandpass shaped color matching functions, *cf.* Figure 2.16, is as follows:

$$C = 1 - R, \quad M = 1 - G, \quad Y = 1 - B \qquad (2.37)$$

In reality, the reflectance curves $\tau(\lambda)$ for the cyan, magenta and yellow inks are far from rectangular (see Figure 2.17) and the relation between CMYK and RGB is not trivial. The problem of obtaining this relation is discussed briefly in Section 2.5.3.3, and we propose an original solution to this problem in Chapter 5.

2.5.3.3   Colorimetric characterization of printers.

The characterization of a color output device such as a digital color printer defines the relationship between the device color space and a device-independent color space, typically based on CIE colorimetry. This relationship defines a (forward) printer model. Several approaches to printer modeling exist in the literature. They may be divided into two main groups, physical and empirical modeling (see *e.g.* Stone *et al.*, 1988, Hardeberg and Schmitt, 1997, Kang, 1997).

■ **Physical models.** Such models are based on knowledge of the physical or chemical behavior of the printing system, and are thus inherently dependent on the technology used (ink jet, dye sublimation, etc.). An important example of physical models for halftone devices is the Neugebauer model, (Neugebauer, 1937, Kang, 1997) which treats the printed color as an additive mixture of the tristimulus values of the paper, the primary colors, and any overlap of primary colors. More recent applications of analytical modeling are illustrated

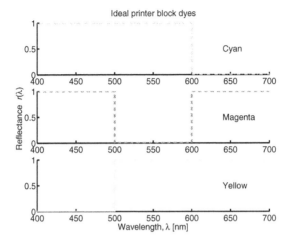

(a) Ideal non-overlapping block inks

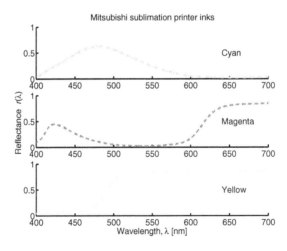

(b) Real sublimation printer inks

Figure 2.17: *Comparison of ideal and real printer inks.*

with a study of Berns (1993b) which applies a modified version of
the Kubelka-Munk spectral model (Kubelka and Munk, 1931) to a
dye diffusion thermal transfer printer.

■ **Empirical models.** Such models do not explicitly require knowledge of the physical properties of the printer as they rely only on
the measurement of a large number of color samples, used either to
optimize a set of linear equations based on regression algorithms,
or to build lookup-tables for 3D interpolation. Regression models
have not been found to be very successful in printer modeling (Hung,
1993), while the lookup-table method is used by several authors, for
example Hung (1993) and Balasubramanian (1994).

However, to be of practical use for image reproduction, these printer
models have to be inverted, and the solution to this problem is rather difficult to find (Iino and Berns, 1998a;b). Iterated optimization algorithms are
often needed to determine the device color coordinates which reproduce a
given color defined in a device-independent color space, as proposed for
example by Mahy and Delabastita (1996).

Another issue that cannot be avoided when discussing printer characterization is gamut mapping (Morovic, 1998). The color gamut of a device
such as a printer is defined as the range of colors that can be reproduced
with this device. Gamut mapping is needed whenever two imaging devices
do not have coincident color gamuts, in particular when a given color in
the original document cannot be reproduced with the printer that is used.
We treat this subject briefly in Appendix E.

We propose in Chapter 5 a novel characterization technique which provides a practical tool to transform any point of the CIELAB space into its
corresponding CMY values (Hardeberg and Schmitt, 1997, Schmitt and
Hardeberg, 1997). This process also includes a color gamut mapping technique, which can be of any type.

2.5.4 Multi-channel imaging

As early as in 1853 Hermann Günter Grassmann stated that three variables
are necessary and sufficient to characterize a color (Section 2.4.1). This
principle, the three-dimensionality of color, has since been confirmed by
thorough biological studies of the human eye. This is the reason why digital color images are composed of three channels or layers, typically red,
green and blue.

However, for digital image acquisition and reproduction, three-channel images have several limitations. First, in a color image acquisition process, the scene of interest is imaged using a given illuminant. Due to metamerism, the color image of this scene under another illuminant cannot be accurately estimated. Furthermore, since the spectral sensitivities of the acquisition device generally differ from the standardized color matching functions, it is also impossible to obtain device-independent color. By increasing the number of channels in the image acquisition device we can remedy these problems, and thus increase the color quality significantly. Several research groups worldwide are working on these matters, for example at the university of Chiba, Japan (Haneishi *et al.*, 1997, Yokoyama *et al.*, 1997, Miyake and Yokoyama, 1998), at Rochester Institute of Technology, USA (Burns and Berns, 1996, Burns, 1997, Berns, 1998, Berns *et al.*, 1998), and at RWTH Aachen, Germany (Keusen, 1996, König and Praefcke, 1998a;b, Hill, 1998). In Chapter 6 we describe our approach to the acquisition of multispectral images with the use of a high definition digital camera and a given number of chromatic filters.

For printing applications more than three image channels have been used for a long time, in particular, a black ink (K) is used in addition to the three subtractive primaries (CMY), as described previously. This has two main advantages, reducing the cost (black can be made with one ink instead of three), and increasing the gamut (more nuances in the dark colors).

Quite recently, desktop printers with six and seven inks have become available. The use of more than four printing inks is often denoted Hi-Fi color, and was up till now only used in very expensive high-end printing systems. Two main methods are used, adding intermediary colors between the subtractive primaries to increase the gamut (and economize), and adding lighter versions of the primary inks, to produce smoother images with less visible dithering. The colorimetric characterization of such printers is an important research field today (see *e.g.* MacDonald *et al.*, 1994, Herron, 1996, Van De Capelle and Meireson, 1997, Mahy and DeBaer, 1997, Berns *et al.*, 1998, Tzeng and Berns, 1998). Another possibility of multi-ink printing is to reproduce not only the wanted color, but the desired spectral reflectance, for example to create a spectral match to an original, and thus avoiding the problems caused by a metameric match, when changing observer or illumination. This is a very new research area (Berns *et al.*, 1998, Tzeng and Berns, 1998).

2.6   Conclusion

In this chapter we have first given our view of the relations and interactions between light, objects, and human color vision. Hopefully this has shed some light on the difficult question concerning what color *really* is. Having defined these basic principles, we have proceeded to a review of different aspects of the science and technology of digital color imaging. Important points are color management, colorimetric characterization of image acquisition and reproduction devices, and imaging using more than three channels.

Chapter 3

Colorimetric scanner characterization

IN THIS CHAPTER, METHODS FOR THE COLORIMETRIC CHARACTERI-
ZATION OF COLOR SCANNERS ARE PROPOSED. THESE METHODS AP-
PLY EQUALLY TO OTHER COLOR IMAGE INPUT DEVICES SUCH AS DIG-
ITAL CAMERAS. THE GOAL OF OUR CHARACTERIZATION IS TO ESTAB-
LISH THE RELATIONSHIP BETWEEN THE DEVICE-DEPENDENT COLOR
SPACE OF THE SCANNER AND THE DEVICE-INDEPENDENT CIELAB
COLOR SPACE. THE SCANNER CHARACTERIZATION IS BASED ON POLY-
NOMIAL REGRESSION TECHNIQUES. SEVERAL REGRESSION SCHEMES
HAVE BEEN TESTED. THE RETAINED METHOD CONSISTS IN APPLYING
A NON-LINEAR CORRECTION TO THE SCANNER RGB VALUES FOL-
LOWED BY A 3RD ORDER 3D POLYNOMIAL REGRESSION FUNCTION
DIRECTLY TO CIELAB SPACE. THIS METHOD GIVES VERY GOOD RE-
SULTS IN TERMS OF RESIDUAL COLOR DIFFERENCES. THIS IS PARTLY
DUE TO THE FACT THAT THE RMS ERROR THAT IS MINIMIZED IN THE
REGRESSION CORRESPONDS TO ΔE_{ab} WHICH IS WELL CORRELATED
TO VISUAL COLOR DIFFERENCES.

3.1   Introduction

To achieve high image quality throughout a digital image system, the first requirement is to ensure the quality of the device that captures real-world physical images to digital images. Several different types of such devices exist, we treat here the case of a flatbed scanner, but the results can also be applied to other devices such as digital cameras. Several factors have influence on this quality, optical resolution, bit depth, spectral sensitivities, noise, to mention a few. In this chapter we will concentrate on the colorimetric faculties of the scanner, that is, the scanner's ability to deliver quantitative device-independent digital information about the colors of the original document. Very few scanners deliver directly colorimetric data, thus we perform a *colorimetric characterization* of the scanner to obtain the relation between the scanner's device dependent RGB color coordinates and a device-independent color space, in our case we use the CIE-LAB pseudo-uniform color space, as defined in Section 2.4.8.1. Several approaches to this characterization exist, as described in Section 2.5.2.1.

For the colorimetrical characterization of a scanner we propose to use an analytical model. The term analytical signifies that it is only based on measurements, no assumption is made about the physical properties of the scanner, as opposed to when using spectral models, see Sections 2.5.2.1 and 6.2. The method is based on polynomial regression and the minimization of ΔE_{ab}, the Euclidean distance in CIELAB space.

In order to characterize the scanner we seek to define the transformation

$$[L^*, a^*, b^*] = g(R, G, B),   \eqno{(3.1)}$$

which converts the RGB scanner components into CIELAB values. Unless the scanner is *colorimetric*, that is, the spectral sensitivities of the three scanner channels equals the CIEXYZ color matching functions or any nonsingular linear transformation of them, an exact analytical representation of Equation 3.1 does not exist.[1] We must thus try to approximate this function. In the literature (see *e.g.* Hung, 1991, Kang, 1992, Wandell and Far-

---

[1] The requirement for *colorimetric* scanners is often referred to as the *Luther-Ives condition* (Ives, 1915, Luther, 1927). More recent work on colorimetric scanner requirements has been done by Hung (1991) and Engeldrum (1993). For image acquisition devices with more than three channels, this requirement can be generalized to requiring that the Human Visual Sub-Space (HVSS, see Section 2.3) be contained in the *sensor visual space* defined as the subspace spanned by the spectral sensitivity functions of the image acquisition device, see Sharma and Trussell (1997a).

rell, 1993, Berns, 1993a, Haneishi *et al.*, 1995, Johnson, 1996), the most
common solution to this problem is to apply linear or higher order regression algorithms to convert from scanner RGB values to CIEXYZ values,
and then apply the formulae given in Section 2.4.8.1 to obtain the CIELAB
values if those are needed. The main drawback with such methods is that
the error that is minimized by the regression algorithm, the RMS error in
CIEXYZ space, is very poorly correlated to visual color differences.

One way to remedy this is to make sure that the output values of the
regression algorithm are CIELAB values, instead of CIEXYZ values, since
the Euclidean distance in CIELAB space corresponds quite well to perceptual color differences. There is clearly not a linear relationship between
scanner RGB and CIELAB space, and we propose thus to model the transformation $g(\cdot)$ given above by nth order polynomials whose coefficients
may be optimized by standard regression techniques (Albert, 1972). In addition to the main step defined by the polynomial regression, we can add
other non-linear transformations steps before and after.

In the following sections we propose several methods for the colorimetric scanner characterization, and we perform a rigorous analysis of their
performance.

3.2   Characterization methodology

The characterization is done as follows, see Figure 3.1. A color chart containing a set of N color samples with known CIELAB values is scanned.
By a picture processing routine we segment each color sample and calculate the mean values of its RGB scanner components. By comparing the
scanned values with the known theoretical CIELAB values for each test
patch, we can find the characterization of the scanner. From this characterization, we will be able to correct the values given by the scanner, to
obtain color consistency, in particular by creating a 3D look-up table for
the RGB-CIELAB transformation that can serve as a *scanner profile* for a
Color Management System (CMS).

Figure 3.2 illustrates the method of approximating the function $g(\cdot)$ by
the function $g'(\cdot)$. For each color $\mathbf{P}_i = [R_i, G_i, B_i]$, $i = 1, \ldots, N$ on the
test chart, the corresponding theoretical values $\mathbf{O}_i^{(t)} = [L_i^{(t)}, a_i^{(t)}, b_i^{(t)}]$ in
CIELAB space are known. The values $\mathbf{O}_i^{(t)}$ have been calculated from the
reflectance spectra of the patches measured by spectrophotometry. Nominal values provided by the color chart manufacturer can also be used, if

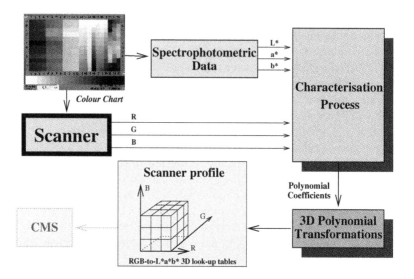

Figure 3.1: *The scanner characterization process, providing a 3D look-up table that can be used as device profile in a Color Management System (CMS) for the conversion of images from scanner RGB to CIELAB.*

we are confident in the quality of the color chart we use, and in the data provided by the manufacturer.

Using these values $\mathbf{O}_i^{(t)}$ and the values $\mathbf{P}_i$ as input to the characterization algorithm, we seek to find the best coefficients of the function $g'(\cdot)$, minimizing the mean square ΔE_{ab} error between all the theoretical values $\mathbf{O}_i^{(t)} = g(\mathbf{P}_i)$ and the approximated ones $\mathbf{O}_i^{(c)} = g'(\mathbf{P}_i)$, as described in the next section and in Appendix A.1.

3.2.1 Regression

The core of our characterization method is in the linear regression step. A linear regression on a vectorial transformation from $\mathbb{R}^3$ to $\mathbb{R}^3$ (such as Equation 3.1) is equivalent to three independent linear regressions on a scalar transformation from $\mathbb{R}^3$ to $\mathbb{R}$ corresponding to each of its components (see Appendix A.1). To simplify the notation, let us consider simply a general transformation

$$y = g(\mathbf{x}), \tag{3.2}$$

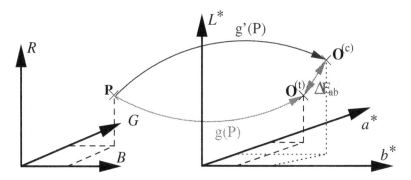

Figure 3.2: *The transformation from scanner RGB-space to CIELAB-space using the function* $g'(\mathbf{P})$. *The difference between the theoretical CIELAB color point* $\mathbf{O}^{(t)} = g(\mathbf{P})$ *and the approximated color* $\mathbf{O}^{(c)} = g'(\mathbf{P})$ *corresponds to the psychophysically relevant color difference* ΔE_{ab}. *The function* $g'(\mathbf{P})$ *is defined by polynomial regression to minimize the RMS* ΔE_{ab} *over the* N *patches of a color chart.*

where $\mathbf{x} \in \mathbb{R}^3$ (RGB space) and $y \in \mathbb{R}$ (one of the components of CIELAB space). We approximate the function $g(\cdot)$ by the following expression,

$$\tilde{y} = g'(\mathbf{x}) = \mathbf{v}^t \mathbf{a}, \qquad (3.3)$$

where the entries of the vector $\mathbf{v}$ are M functions $h_i(\mathbf{x})$ of the input values,

$$\mathbf{v} = [h_0(\mathbf{x}), h_q(\mathbf{x}), \dots, h_{M-1}(\mathbf{x})]^t, \qquad (3.4)$$

and $\mathbf{a} = [a_0, a_1, \dots, a_M]^t$ is a vector of coefficients to be optimized. For the simple example of linear regression with a first order polynomial, $M = 4$,

$$\mathbf{v} = [1, R, G, B]^t,$$

and Equation 3.3 becomes simply $\tilde{y} = a_0 + a_1 R + a_2 G + a_3 B$. For the example of a 3rd order polynomial with all the cross-product terms, $M = 20$, and

$$\begin{aligned}
\mathbf{v} =\ & [1,\ R,\ G,\ B,\ R^2,\ RG,\ RB,\ G^2,\ GB,\ B^2, \\
& R^3,\ R^2G,\ R^2B,\ RG^2,\ RGB, \\
& RB^2,\ G^3,\ G^2B,\ GB^2,\ B^3]^t.
\end{aligned} \qquad (3.5)$$

Given *i)* a set of input data $\mathbf{x}_j$, $j = 1, \ldots N$, *ii)* their corresponding vectors $\mathbf{v}_j$, and *iii)* the observed output data y_j, then the coefficient vector **a** which minimizes the RMS difference between observed and predicted data, is given by (see Appendix A.1)

$$\mathbf{a} = (\mathbf{V}^t\mathbf{V})^{-1}\mathbf{V}^t\mathbf{y} = \mathbf{V}^-\mathbf{y}, \qquad (3.6)$$

where $\mathbf{V} = [\mathbf{v}_1\ \mathbf{v}_2\ \ldots \mathbf{v}_N]^t$, $\mathbf{y} = [y_1\ y_2\ \ldots\ y_N]^t$, and $\mathbf{V}^-$ is the Moore-Penrose pseudo-inverse of $\mathbf{V}$ (Albert, 1972).

A very important factor concerning the success of a regression algorithm is the choice of the function $h_i(\cdot)$ defining the vectors $\mathbf{v}$, so that the regression function $g'(\mathbf{x})$ provides a good approximation of $g(\mathbf{x})$. Typically, for mth order polynomial regression, if we choose m too low, $g'(\mathbf{x})$ will not have enough *degrees of freedom* to "follow" $g(\mathbf{x})$, while if m is chosen too large, $g'(\mathbf{x})$ can tend to oscillate, see Figure 3.3. An important step for the choice of $h_i(\cdot)$ is the linearization of the scanner RGB values, as described in the next section.

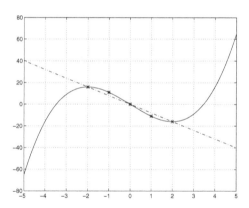

Figure 3.3: *1D example of first and third order polynomial regression functions* $\tilde{y} = g'(x)$ *applied to a data set* $y_i = g(x_i)$, $i = 1, \ldots, 5$ *marked with stars* (∗). *The third order polynomial function* (—) *gives zero residual error, but nevertheless it is obviously less adapted than the first order function* (− · −) *outside of the domain.*

3.2.2   Linearization of the scanner RGB values

The CCD scanner is inherently a linear electro-optic conversion device, that is, it converts the optical energy of the incoming light into proportional amounts of electric signals, see Section 6.2.1. These signals are in turn discretized and presented as digital k-bit RGB values at the output (typically $k = 8$ for a low-cost scanner, 10 or 12 for more professional ones). However, we have observed in practice that the scanner RGB values are often not proportional to the spectral energy. This non-linearity may have several causes, and we mention some.

1. Black offset. Even in total absence of incident light, the CCD sensor produces a small electric signal due to leak currents.

2. Deliberate corrections to enhance the quality of the display on a computer monitor by counteracting the non-linear transfer function of the monitor.[2] Such corrections are often called gamma-corrections (Poynton, 1996), and may also be applied to minimize the noise due to quantization. For a given scanner with its scanner driver software, the parameters of this correction may or may not be known to the user.

3. Stray light in the acquisition system may cause image-dependent deviation from linearity.

4. Fluorescence of the scanned reflective media causes the linear model of the scanner to fail.

5. Limited dynamic range of the detector.

6. Inclusion of ultraviolet and infrared radiation in the measurements.

In general, the user has limited knowledge of these factors, and we proceed thus to an automatic characterization of the linearity and eventually to a linearization of the scanner RGB values.

---

[2] The intensity of light generated by a display device is not usually a linear function of the applied signal. A conventional CRT has a power-law response to voltage: light intensity produced at the face of the display is approximately the applied voltage, raised to the 2.5 power. Gamma correction is the process of compensating for this non-linearity by transforming linear scanner values to a nonlinear video signal by a power-law function (Poynton, 1996).

What we wish to achieve through the linearization[3] of the scanner RGB values is to obtain RGB values that are proportional to the optical energy of the input light, as illustrated in Figure 3.4. This correction is called *gray balance* by some authors (Kang, 1992; 1997)

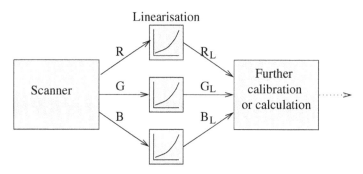

Figure 3.4: *Linearization of the scanner output values. The scanner device coordinates* (R, G, B) *are corrected to obtain values* (R_L, G_L, B_L) *that are proportional to the optical energy of the incident light.*

Two different approaches for the linearization are presented in the following sections, a global and a piecewise linear approach. They both rely on the comparison of measured *reflectance factors* r_i, $i = 1, \ldots, N_g$ of each of the N_g patches of a grayscale color chart with the corresponding mean R_i, G_i and B_i device coordinates from the scanner.

3.2.2.1 A global approach

Using this approach, we assume that the non-linearity of the scanner stems mainly from a CRT gamma-correction (Berns *et al.*, 1993, Poynton, 1996), thus there is a power-law relation between the optical and electrical values, as given by

$$O = k(e - e_d)^\gamma \tag{3.7}$$

Here, we use O to indicate the optical signal, e the electrical signal, e_d the electrical signal corresponding to dark current, k a scaling factor and γ the exponent. If we denote the linearized RGB values as R_L, G_L and B_L we

---

[3]What wee seek to do is to "remove" the gamma correction that was imposed by the scanner, *cf.* item 2 above.

have the following three equations.

$$
\begin{aligned}
R_L &= k_R(R - e_{d_R})^{\gamma_R} \\
G_L &= k_G(G - e_{d_G})^{\gamma_G} \\
B_L &= k_B(B - e_{d_B})^{\gamma_B}
\end{aligned}
\tag{3.8}
$$

To perform the linearization we then need to estimate the unknown parameters in these equations. The dark current values, which are the device coordinates resulting from a black object, may be measured experimentally. For the Sharp JX-300 flatbed scanner we obtained the following mean values simply by performing a scan, with an open cover, without any document, and in a completely dark room:

$$
e_{d_R} = 7.05 \qquad e_{d_G} = 3.80 \qquad e_{d_B} = 6.10
$$

To estimate the gamma values, we would like the linearized values to be equal to the measured reflectance factors r_i, $i = 1, \ldots, N_g$, for the N_g grayscale chart patches. We then have the following set of equations.

$$
\begin{aligned}
r_i &= k_R(R_i - e_{d_R})^{\gamma_R}, \\
r_i &= k_G(G_i - e_{d_G})^{\gamma_G}, \\
r_i &= k_B(B_i - e_{d_B})^{\gamma_B}, \qquad i = 1, \ldots, N_g
\end{aligned}
\tag{3.9}
$$

To solve this for $\gamma_R, \gamma_G, \gamma_B$, we take the logarithm on both sides of the equations.

$$
\begin{aligned}
\log(r_i) &= \log k_R + \gamma_R \log(R_i - e_{d_R}) \\
\log(r_i) &= \log k_G + \gamma_G \log(G_i - e_{d_G}) \\
\log(r_i) &= \log k_B + \gamma_B \log(B_i - e_{d_B}) \qquad i = 1, \ldots, N_g
\end{aligned}
\tag{3.10}
$$

This set of N_g equations can easily be solved by a least mean squares approach. We note that with this method, we do not get an exact match between the reflectances and the linearized grayscale values, as can be seen in Figure 3.5.

3.2.2.2 A piecewise linear approach

Using this approach, we consider the linearization curve to be piecewise linear in the *semilog* space. By ordering the reflectance values such that

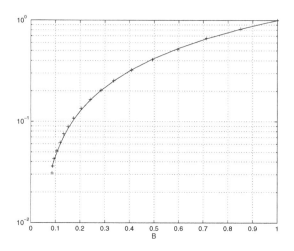

Figure 3.5: *Linearization curve with a global approach (—) compared with the measured reflectance values (+). We see that the linearization curve differs from the reflectance values, especially at low levels. For clarity we have only included the result for the blue (B) channel.*

$r_i < r_{i+1}$, $i = 1, \ldots, N_g$ (see Figure 3.6) we obtain the following set of $N_g - 1$ equations for the R channel (similarly for the G and B channels):

$$\log R_L = a_i R + b_i, \quad R_i < R \leq R_{i+1}, \quad 1 \leq i \leq N_g - 1 \quad (3.11)$$

The coefficients a_i and b_i, $i = 1, \ldots, N_g$ of Equation 3.11 are calculated as follows:

$$a_i = \frac{\log r_{i+1} - \log r_i}{R_{i+1} - R_i}, \quad b_i = \frac{R_{i+1} \log r_i - R_i \log r_{i+1}}{R_{i+1} - R_i} \quad (3.12)$$

For input values $R > R_{N_g}$ we simply perform an extrapolation of the $(N_g - 1)$'th segment. For values $R < R_1$ we perform a linear interpolation in linear space between the point (R_1, r_1) and $(e_{d_R}, 0)$. We cannot do this interpolation in semilog space, as we need to reach zero.

This method guarantees that on the grayscale, the linearized values will be exactly equal to the reflectance values. Furthermore, it assures proper handling of the input values near zero. The resulting linearization curves for R, G, and B are shown in Figure 3.7.

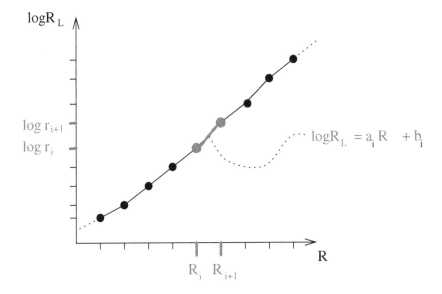

Figure 3.6: *Illustration of a piecewise linear linearization curve in semilog space. This approach guarantees that the linearized values equal the measured reflectance values on the points of the grayscale chart.*

3.2.2.3 Testing of the linearization algorithms

Linearization of the RGB scanner output have been done for several different scanners and digital cameras and several different gray-scale color charts. The preferred method depends on the application and on the device. The quality of the proposed linearization methods is not easy to evaluate, since the linearization is normally only the first step of a series of correction algorithms, of which the final outcome depend.

We have tested several combinations of linearization methods (piecewise linear, global, none) together with following polynomial corrections for the Sharp JX-300 24 bit/pixel flatbed color scanner and an AGFA photographic grayscale color chart with $N_g = 18$ color patches. The graphs and numbers presented in this section is from this experiment which is explained in detail in (Hardeberg, 1995). No clear conclusions are drawn, however the tendency is that the piecewise linear method gives the best results.

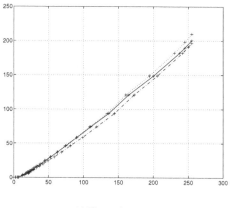

(a) The entire range.

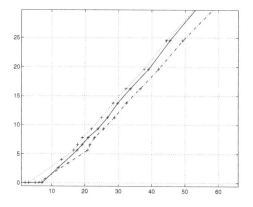

(b) A zoom of the region around the origin.

Figure 3.7: *The linearization curves for R (—), G (· · ·) and B (− · −) obtained with a piecewise linear approach. The ordinate axes represent the input values (R, G, B), the abscissa axes represent the linearized values (R_L, G_L, B_L). We see that we get an exact match between the reflectances marked with crosses and the linearized grayscale values. We also note the behavior near the origin, where special considerations have been taken.*

For the experiments with the AGFA Arcus II 36 bits/channel flatbed color scanner (see Section 3.3.2) no linearization was needed, while to obtain the very high accuracy needed for the multispectral experiments of Chapter 7 with the PCO SensiCam digital camera, a correction for the black offset was needed.

Note that other linearization schemes could have been used, such as first, second or third order polynomial regression, piecewise linear interpolation in linear or log-log space, or spline interpolation.

In the remaining of this chapter we will assume that the scanner output has been linearized if this has been found necessary.

3.2.3   Choice of the approximation function

Crucial for the performance of the characterization is the choice of the function to approximate $g(\cdot)$. We have tested a great number of different implementations, using different color charts, different polynomial orders, including or not linearization and a preliminary power-law correction, etc. Preliminary results can be found in (Schmitt *et al.*, 1990; 1995; 1996, Hardeberg, 1995, Hardeberg *et al.*, 1996). Here we present some of the proposed methods. To obtain a better understanding of the successive steps of the methods, we illustrate them with symbolic equations.

3.2.3.1   Linear regression to XYZ space

With this classical method (Hung, 1991, Kang, 1992, Wandell and Farrell, 1993, Berns, 1993a, Haneishi *et al.*, 1995, Johnson, 1996), a linear regression algorithm 'T1' (without the constant term, that is, $a_0 = 0$) is applied to convert from scanner RGB values to CIEXYZ values, and then the standardized formula (labeled 'CIE') given in Section 2.4.8.1 is applied to obtain the CIELAB values if those are needed. Practically, the XYZ values are obtained by multiplying the RGB values by a 3x3 matrix of parameters.

$$(T1, XYZ): \quad \begin{bmatrix} R \\ G \\ B \end{bmatrix} \overset{T1}{\Rightarrow} \begin{bmatrix} X \\ Y \\ Z \end{bmatrix} \overset{CIE}{\Rightarrow} \begin{bmatrix} L^* \\ a^* \\ b^* \end{bmatrix} \qquad (3.13)$$

3.2.3.2 Second order polynomial regression to XYZ space

Here, we hope to obtain a better fit by applying a second order polynomial regression algorithm 'T2' to convert from scanner RGB values to CIEXYZ values. This corresponds to a 3x10 correction matrix when we include a constant term.

$$(\text{T2, XYZ}): \quad \begin{bmatrix} R \\ G \\ B \end{bmatrix} \stackrel{\text{T2}}{\Rightarrow} \begin{bmatrix} X \\ Y \\ Z \end{bmatrix} \stackrel{\text{CIE}}{\Rightarrow} \begin{bmatrix} L^* \\ a^* \\ b^* \end{bmatrix} \qquad (3.14)$$

3.2.3.3 Third order polynomial regression to XYZ space

With this method, a third order polynomial regression algorithm 'T3' is applied to convert from scanner RGB values to CIEXYZ values, corresponding to a 3x20 correction matrix.

$$(\text{T3, XYZ}): \quad \begin{bmatrix} R \\ G \\ B \end{bmatrix} \stackrel{\text{T3}}{\Rightarrow} \begin{bmatrix} X \\ Y \\ Z \end{bmatrix} \stackrel{\text{CIE}}{\Rightarrow} \begin{bmatrix} L^* \\ a^* \\ b^* \end{bmatrix} \qquad (3.15)$$

3.2.3.4 Polynomial regression to CIELAB space

The main drawback with all the methods presented up to now is that the error that is minimized by the regression algorithm, the RMS error in CIEXYZ space, is very poorly correlated to visual color differences. We propose thus a regression scheme in which the output values of the regression algorithm are CIELAB values, instead of CIEXYZ values, since the Euclidean distance in CIELAB space corresponds quite well to perceptual color differences. There is clearly not a linear relationship between scanner RGB and CIELAB space, and we propose thus to model the transformation $g(\cdot)$ of Equation 3.1 directly by nth order polynomial regression, $n = 1, 2, 3$.

$$(\text{T}n, \text{XYZ}): \quad \begin{bmatrix} R \\ G \\ B \end{bmatrix} \stackrel{\text{T}n}{\Rightarrow} \begin{bmatrix} L^* \\ a^* \\ b^* \end{bmatrix} \qquad (3.16)$$

3.2.3.5 Non-linear correction followed by polynomial regression

With this method, we have applied a non-linear correction of the RGB values before the regression by applying the cubic root function, *i.e.* the functions $h_i(R, G, B)$ (cf. Eq.3.4) are replaced by $h_i(R^{1/3}, G^{1/3}, B^{1/3})$. The use of this cubic root function is motivated from considering the CIELAB transformations, which involves such cubic root functions on the XYZ tristimulus values, which again are proportional to the optical energy.

$$(\text{p=1/3, T}n\text{, LAB)}: \quad \begin{bmatrix} R \\ G \\ B \end{bmatrix} \overset{p=\frac{1}{3}}{\Rightarrow} \begin{bmatrix} R^{\frac{1}{3}} \\ G^{\frac{1}{3}} \\ B^{\frac{1}{3}} \end{bmatrix} \overset{\text{T}n}{\Rightarrow} \begin{bmatrix} L^* \\ a^* \\ b^* \end{bmatrix} \quad (3.17)$$

where '$p = \frac{1}{3}$' indicates that the components are raised to the power of one third.

3.3 Experimental results

We have successfully applied the described colorimetric characterization algorithms to several different scanners and also digital cameras. Some of these results are reported in other chapters of this dissertation. We will here report the results with the AGFA Arcus II scanner on which most of the experimentation has been effectuated.

3.3.1 Evaluation measures

To be able to evaluate and compare the different approaches, the following measures are provided:

- $\overline{\Delta E}$ which is the mean ΔE_{ab} color difference (Section 2.4.8.1) between the calculated and theoretical CIELAB-values for the complete test chart.

- ΔE_{max} which represents the maximum ΔE_{ab} color difference between the calculated and the theoretical CIELAB-values.

- $\overline{\Delta L}, \overline{\Delta a}, \overline{\Delta b}, \Delta L_{\text{max}}, \Delta a_{\text{max}}, \Delta b_{\text{max}}$, which are the mean and maximal absolute errors measured on each channel separately.

- σ_E which is the standard deviation of the ΔE_{ab} error over the patches, that is, $\sigma_E^2 = \frac{1}{N} \sum_{i=1}^{N-1} (\Delta E_i - \overline{\Delta E})^2$

3.3.2 Results

We have applied the described characterization methods to the AGFA Arcus II flatbed scanner, using two different IT8.7/2 color charts (Ohta, 1993, ANSI IT8.7/2, 1993) containing $N = 288$ color patches, one from AGFA, with nominal CIELAB values provided by the manufacturer, and another from FUJI, being calibrated, that is, provided with CIELAB values measured on this copy of the chart. We have tested first, second and third order polynomial regression from scanner RGB to XYZ and CIELAB, as well as 1-3rd order regression between the square root of the RGB values and CIELAB space, as described in Section 3.2.3. For the AGFA chart we obtain the results given in Table 3.1, and for the FUJI chart we obtain the results given in Table 3.2.

From these results, several comments may be made. First, we note that mean errors always get smaller as higher order regression is used, as expected. However, maximum errors sometimes get worse. For the regressions to XYZ space, we see that errors on the L^* component are quite small compared to on a^* and b^*. Polynomial regression directly from linear RGB values to CIELAB is clearly not a good solution, while applying a pre-correction that "mimics" the non-linear function involved in the XYZ-CIELAB conversion gives very good results, especially when third order polynomials are used, giving mean residual error of about one ΔE unit. When relating these results to the rule of thumb described in Table 2.1 on page 31, we see that our results are very good. The mean error is hardly perceptible, while the maximal error is perceptible, but acceptable.

Also when comparing to results found in the literature, our results are excellent. For example Haneishi *et al.* (1995) obtained mean/max ΔE_{ab} errors of 4.9/16.6 and 2.0/14.0 using respectively first and second order polynomials to transform from scanner RGB to XYZ for a Canon CLC500 scanner using a chart of 125 patches. Rao (1998) obtained mean/max ΔE errors[4] of 2.33/11.95 using linear least squares from RGB to XYZ, and 1.62/4.55 using a non-linear least squares method minimizing ΔE by a Levenberg-Marquardt iterative optimization scheme. This was done for a IBM TDI/Pro 3000 scanner using the Kodak Q60 IT8 chart. Kang (1992) obtained mean ΔE_{ab} errors of 2.52 and 1.85, using 3×6 and 3×14 matrices, respectively, for the RGB-XYZ conversion for a Sharp JX-450

---

[4]Rao (1998) define a somewhat peculiar $\Delta E = \sqrt{\Delta L^{*2}/4 + \Delta a^{*2} + \Delta b^{*2}}$. This is clearly lowering their ΔE values compared to the standardized $\Delta E_{ab} = \sqrt{\Delta L^{*2} + \Delta a^{*2} + \Delta b^{*2}}$ used in our work.

Table 3.1: *Results of the characterization methods for the AGFA Arcus II scanner, with the AGFA IT8.7/2 color chart.*

Method	$\overline{\Delta E}$	ΔE_{max}	$\overline{\Delta L}$	ΔL_{max}	$\overline{\Delta a}$	Δa_{max}	$\overline{\Delta b}$	Δb_{max}	σ_E
T1, XYZ	4.841	22.939	1.276	3.715	2.782	20.932	3.135	15.354	3.800
T2, XYZ	2.989	28.246	0.458	3.585	1.653	21.718	1.956	28.089	3.811
T3, XYZ	2.170	20.903	0.427	2.490	1.306	18.576	1.348	13.276	2.772
T1, LAB	22.27	49.111	15.31	34.052	8.422	37.251	9.056	40.062	9.298
T2, LAB	8.858	40.349	2.854	14.054	5.077	25.858	5.024	29.930	5.682
T3, LAB	5.386	30.792	1.385	9.775	3.207	23.060	3.114	19.784	4.065
p=1/3, T1, LAB	5.652	23.961	3.241	11.345	2.234	23.304	2.987	12.645	3.339
p=1/3, T2, LAB	1.496	12.448	0.348	2.352	1.166	12.348	0.579	3.311	1.341
p=1/3, T3, LAB	**0.918**	**4.666**	**0.289**	**2.069**	**0.621**	**4.588**	**0.427**	**2.792**	**0.658**

Table 3.2: *Results of the different characterization methods for the AGFA Arcus II scanner, with the FUJI IT8.7/2 color chart.*

Method	$\overline{\Delta E}$	ΔE_{max}	$\overline{\Delta L}$	ΔL_{max}	$\overline{\Delta a}$	Δa_{max}	$\overline{\Delta b}$	Δb_{max}	σ_E
T1, XYZ	5.079	17.658	1.676	5.162	2.925	14.157	3.379	11.970	3.868
T2, XYZ	2.145	18.457	0.554	6.136	1.303	16.672	1.243	8.273	2.557
T3, XYZ	1.574	15.046	0.468	5.008	0.955	13.270	0.898	7.128	2.013
T1, LAB	19.38	45.269	13.63	35.089	7.719	32.522	7.450	35.796	9.486
T2, LAB	7.288	36.194	2.504	15.418	4.282	24.528	3.979	25.483	5.305
T3, LAB	4.563	25.965	1.294	10.909	2.673	20.541	2.633	16.124	3.763
p=1/3, T1, LAB	5.462	18.393	2.352	11.748	2.509	17.901	3.640	12.083	3.602
p=1/3, T2, LAB	1.350	8.503	0.486	3.321	0.953	8.448	0.548	4.413	1.108
p=1/3, T3, LAB	**1.006**	**5.515**	**0.432**	**2.430**	**0.606**	**3.649**	**0.488**	**3.644**	**0.681**

scanner using a Kodak Q60-C IT8 color chart.

To gain more insight into the performance of the proposed methods, we present in Figure 3.8 histograms of the residual ΔE_{ab} errors, in Figure 3.9 the distribution of errors versus lightness, and in Figures 3.10 and 3.11, graphical visualizations of the errors in CIELAB space. For a more detailed presentation of the numerical data involved in the characterization process, refer to Appendix C.

3.3.3   Generalization

An important question concerning our approach is the generalization. Would the polynomial fit those colors that are not used in the regression? With which colorimetric errors? How many color patches are needed in the regression to warrant an acceptable error? To answer these questions we have conducted the following experiment. Using the AGFA IT8.7/2 color chart which has $N = 288$ patches, we systematically reduce the number of patches used for the regression $N_{training}$, and use the $N_{testing} = N - N_{training}$ removed patches for testing. The results, using the best method (p=1/3, T3, LAB) are reported in Table 3.3. We see that even when reducing the training set to 54, the mean and max total error only increases by 0.48 and 1.45 ΔE units, respectively. When the training set contains fewer patches than the number of parameters (20) the system gets underdetermined and the estimation is unusable. With a sufficient number of patches, approximatively the half, the error on the testing set is only slightly higher than on the training set. These results are in good agreement with what was found for lower order polynomials by Kang (1992).

3.3.4   Comparison of results with and without characterization

To evaluate the results of our characterization process, it would be interesting to compare the error in terms of ΔE_{ab} with and without characterization, that is, for each patch of the color chart, to compare the CIELAB-values obtained by our characterization with the CIELAB-values obtained *without* characterization. This is, however, not a straightforward task, the problem being how to obtain "good" *uncharacterized* CIELAB-values.

In the absence of a characterization procedure, a natural approach would be to follow what a typical user probably would do, by first adjusting the gamma corrections at the scanning step in order to choose the preferred

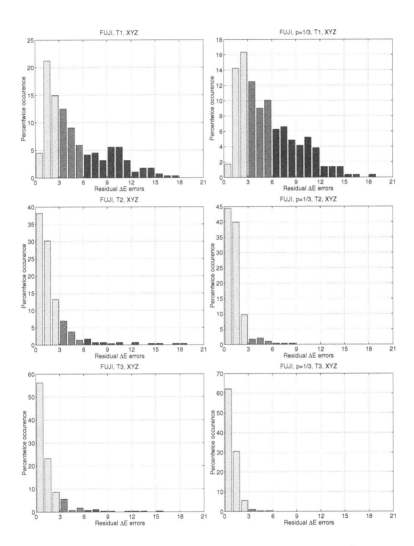

Figure 3.8: *Error histograms for the FUJI IT8.7/2 chart using first, second and third order regression to XYZ space (left), and to CIELAB space, including a cubic root pre-correction function (right).*

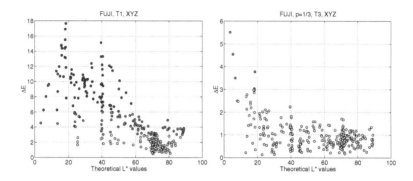

Figure 3.9: *Error distribution for the FUJI IT8.7/2 chart using traditional linear regression (T1, XYZ) (left) and our preferred method (p=1/3, T3, LAB) (right). (Notice the scale difference.)*

Table 3.3: *Results for the (p=1/3, T3, LAB) method and the AGFA IT8.7/2 color chart using different numbers of patches for training and testing. We see that the overall error increases only slightly even when the number of patches used for training is reduced to 25% of the total number of patches. However, the errors increase rather quickly if the number of patches is reduced even further.*

# Patches		Training		Testing		Total	
N_{training}	N_{testing}	$\overline{\Delta E}$	$\Delta E_{\max}$	$\overline{\Delta E}$	$\Delta E_{\max}$	$\overline{\Delta E}$	$\Delta E_{\max}$
288	0	0.918	4.666	—	—	0.918	4.666
270	18	0.901	4.483	1.164	3.958	0.918	4.483
252	36	0.897	4.437	1.062	3.881	0.918	4.437
216	72	0.896	4.037	1.105	3.757	0.948	4.037
180	108	0.940	3.945	0.982	3.629	0.956	3.945
144	144	0.978	3.931	0.961	3.358	0.969	3.931
108	180	0.956	3.818	1.019	4.764	0.995	4.764
72	216	0.929	3.341	1.060	7.171	1.027	7.171
63	225	0.937	3.208	1.145	6.032	1.100	6.032
54	234	0.901	2.820	1.249	6.118	1.184	6.118
45	243	0.744	2.245	1.515	15.009	1.394	15.009
36	252	0.720	2.179	1.916	19.808	1.766	19.808
27	261	0.329	0.881	4.048	33.273	3.699	33.273
18	270	0.000	0.000	212.5	2104.9	199.2	2104.9

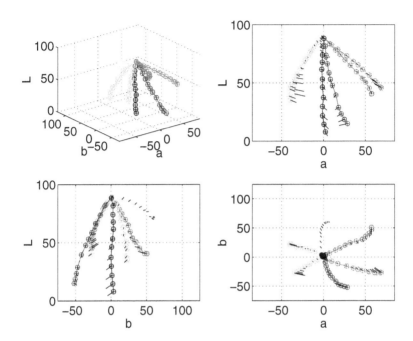

Figure 3.10: *Visualization of the difference between the measured and the estimated CIELAB values of some of the patches of the FUJI chart using the linear regression method (T1, XYZ). The residual color differences appear as black spikes.*

image displayed on a CRT monitor and then by considering the displayed image as it appears on the screen as a reference. With this in mind, it makes sense to define the *uncharacterized* transformation from gamma corrected RGB scanner to CIELAB as the transformation between the RGB space of the CRT monitor and CIELAB.

For the uncharacterized transformation we chose a gamma correction of 2.2 at the scanner step, and then applied a classical CRT model (Berns *et al.*, 1993), this giving a mean error of $\overline{\Delta E} = 9.18$, and a maximum error of $\Delta E_{\max} = 26.1$, which have to be compared to $\overline{\Delta E} = 1.01$ and $\Delta E_{\max} = 5.52$ obtained with our characterization algorithm.

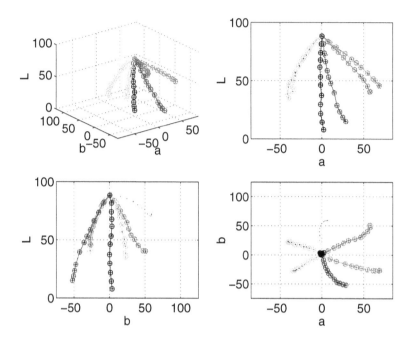

Figure 3.11: *Visualization of the difference between the measured and the estimated CIELAB values of some of the patches of the FUJI chart using the proposed characterization method (p=1/3, T3, LAB).*

3.4   Conclusion

With the methods described in this chapter, a link between the device-dependent color space of a scanner and the device-independent CIELAB color space standardized by the CIE is provided.

The proposed characterization method consists of three steps: *i)* a linearization of the RGB scanner values, *ii)* a preprocessing with a cubic root function, and *iii)* a third order 3D regression polynomial. The preprocessing step serves as a first approximation of the cubic root function involved in the conversion from CIEXYZ to CIELAB.

Applying this method to an AGFA Arcus II flatbed scanner, we obtained a mean error of $\Delta E_{ab} = 0.918$, and a maximum error of $\Delta E_{ab} = 4.666$ between the computed and the measured CIELAB values on the complete set of patches of the AGFA IT8.7/2 color chart. These results are very satisfactory, compared to results obtained in the literature. The algorithms have also been applied to other image acquisition devices, and we conclude that the characterization process introduces a significant improvement of the colorimetric quality of the image acquisition device.

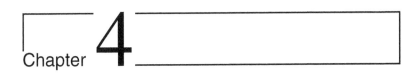

Chapter 4

High quality image capture

IN THE CONTEXT OF ARCHIVING OF FINE ART PAINTINGS, WE HAVE DEVELOPED VARIOUS TECHNIQUES FOR THE DIGITAL ACQUISITION AND THE PROCESSING OF HIGH QUALITY AND HIGH DEFINITION CO-LOR IMAGES. AFTER A SHORT REVIEW OF EXISTING SYSTEMS, WE DE-SCRIBE HERE BRIEFLY THE SUCCESSIVE STEPS OF OUR APPROACH TO THIS PROCESS. AFTER A GENERAL SET UP, WE FIRST RECORD A SET OF EXPERIMENTAL DATA CORRESPONDING TO THE CALIBRATION OF THE CCD ARRAY OF THE DIGITAL CAMERA AND TO THE LIGHT DISTRIBU-TION. THE DIGITAL IMAGE OF THE PAINTING IS THEN SUCCESSIVELY CORRECTED FOR LIGHT DISTRIBUTION INHOMOGENEITIES, CHROMA-TIC ABERRATIONS BY A PRECISE REGISTRATION OF THE THREE CHAN-NELS, AND POOR COLORIMETRIC QUALITY OF THE SPECTRAL RESP-ONSES OF THE DIGITAL CAMERA BY A NON-LINEAR COLORIMETRIC 3D TRANSFORMATION OPTIMIZED USING AN IT8.7/2 COLOR TARGET. THE CORRECTED IMAGE IS THEN DESCRIBED BY DEVICE-INDEPEN-DENT COLOR COMPONENTS AND CAN BE ARCHIVED OR FURTHER CON-VERTED FOR VISUALIZATION ON A CALIBRATED DISPLAY OR PRINTER. FOR VERY HIGH DEFINITION AN IMAGE MOSAICING IS FURTHER PER-FORMED.

4.1 Introduction

Traditionally, the image acquisition process using classical photographic techniques is quite complex and it is difficult to control how color is being processed in the different steps, see Figure 4.1. With the arrival of high-

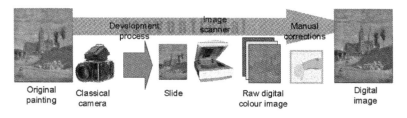

Original   Classical   Slide   Raw digital   Digital
painting   camera   colour image   image

Figure 4.1: *Traditional image acquisition process using classical photography.*

quality high-resolution electronic cameras new possibilities have emerged. The proposed digital image acquisition process is performed directly from the painting, without a photographic intermediary; using a high resolution CCD camera, see Figure 4.2. The original methods that are described in this chapter allow a perfect spatial resolution and excellent color fidelity. The acquisition is therefore independent of the light source and the acquisition equipment. The process of colorimetric characterization of the camera provides the transformation from the RGB values of the camera to the device-independent CIELAB color space, using spectrally calibrated color targets.

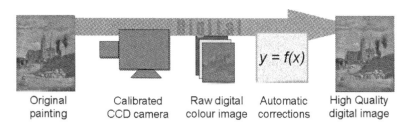

Original   Calibrated   Raw digital   Automatic   High Quality
painting   CCD camera   colour image   corrections   digital image

Figure 4.2: *Entirely digital image acquisition process using a CCD camera.*

In the framework of the European VASARI project[1] the Ecole Nationale Supérieure des Télécommunications (ENST) developed in 1990–1991 a set of techniques for the direct digital acquisition of a painting with a digital camera. Twenty paintings belonging to the Louvre collection was digitized in high resolution and high quality. The experiments were conducted at the French Museum Research Laboratory (LRMF) for obvious security reasons and with the valuable collaboration of their experts. The aim of the VASARI digitization program at the Louvre Museum was to obtain digital acquisitions of the best possible quality under the constraint of physical and technological limitations, so that a painting could be analyzed, displayed and reproduced, in part or totally, and so that it may be studied at several stages of its life.

More recently the ENST and the LRMF again collaborated on the occasion of the exhibition on Jean-Baptiste Camille Corot (1796-1876) in Grand Palais, Paris (January-May 1996). Eight paintings were digitized with an improved color calibration technique and stored on a CD-ROM, which was published and presented during the exhibition (Crettez *et al.*, 1996, Schmitt, 1996, Maître *et al.*, 1997). On later occasions we also participated in the digitization of paintings by Francisco de Goya (1746-1828) and Georges de La Tour (1593-1652) (Crettez, 1998, Hardeberg and Crettez, 1998, Crettez and Hardeberg, 1999)

In the following sections we describe the various operations developed to obtain high resolution and high quality color images in the museum context. Several people have participated in this work, and significant research and development efforts have been done as student projects (Müller, 1989, Camus-Abonneau and Camus, 1989, Allain, 1989, Goulam-Ally, 1990a;b, Deconinck, 1990, Bournay, 1991, Wu *et al.*, 1991, Nagel, 1993). These works have provided indispensable tools and background material for the work of this dissertation.

After giving a review of some existing high resolution digital cameras in Section 4.2, we present in Section 4.3 the general experimental setup and initial calibration of the digital camera including the lighting conditions. Then, in Section 4.4, we describe the three transformations that are successively applied on the digital images recorded directly from a painting: the light distribution homogenization, the inter-channel registration and the colorimetric correction. Finally, different post-processing algo-

---

[1]European ESPRIT II project n° 2649, Visual Arts System for Archiving and Retrieval of Images

rithms are presented briefly in Section 4.5: mosaicing, visualization and reproduction, and colorimetric analysis of paintings.

4.2   High resolution digital cameras, a review

High-end digital cameras is a field of research and development in very rapid development. What was considered as high-end five years ago is generally obsolete today. In this section we make no attempt to give a complete survey of the past, present, and future of high resolution digital cameras. However, we will describe shortly a few examples of such, in particular those developed in the framework of the European ESPRIT projects, VASARI and MARC[2].

4.2.1   The VASARI project

Martinez *et al.* (1993) present the seven-channel VASARI image acquisition system implemented at the National Gallery in London. The system consists of a 3000 × 2300 pixel camera, the Kontron ProgRes 3000 (Lenz and Lenz, 1989), mounted on a repositioning system. By mosaicing they attain a resolution of about $20k \times 20k$. A high-quality lens is chosen having low geometric and radiometric distortion, thus avoiding expensive and inaccurate correction of geometric distortion.

A lighting system with optic light guides that move with the camera is used. In this way the same light-distribution correction can be applied for each sub-image. Furthermore the seven interference filters are introduced between the light source and the painting, not between the painting and the camera. This is intended to reduce misalignment errors as well as reducing the exposure of the painting.

The presented system uses only 8 bit per channel, but an extension to twelve bit is reported to be under investigation.

The seven filters are chosen as broad-band, nearly Gaussian filters with transmittances covering the visible spectrum with considerable overlap. The authors refer to Deconinck (1990) where 12 narrow-band filters are used, but claim that the use of seven instead of twelve filters represents a marginal loss in quality, but a significant gain in processing time.

The conversion from the seven camera responses to CIE XYZ color space is performed using linear regression optimized on the Macbeth Col-

---

[2]European ESPRIT III project, Methodology for Art Reproduction in Colour.

orChecker chart comprising only 24 color patches. The result of the calibration is evaluated on the same chart giving an average error of $\Delta E^*_{CMC} = 2.3$.

Refer to Cupitt (1996) for a practical summary for non-experts of the color camera calibration experience gained at the National Gallery.

Other implementations of VASARI image capture systems can be found *e.g.* at the Neue Pinakothek in Munich (Müller and Burmeister, 1993) and at the University of Firenze (Abrardo *et al.*, 1996).

4.2.2 Further developments under the MARC project

MacDonald and Lenz (1994) present an ultra-high resolution digital camera, developed under the MARC project. Two techniques for attaining high resolution in digital cameras are explained in this paper, micro- and macro-scanning.

In a micro-scanned array camera the image is formed by micro-scanning the intermediate grid positions of a low resolution 2D CCD array (Lenz and Lenz, 1989). The technique is known as piezo-controlled aperture displacement (PAD). The final image is constructed by interlacing the sub-images. The macro-scanning technique consists of shifting the CCD array repeatedly by its width and height, and constructing the final image by mosaicing. MacDonald and Lenz point out that it is convenient to combine these two techniques, and describe two different implementations of this:

- A micro-scanning camera is moved as a whole in front of the object. This is the implementation used by Martinez *et al.* (1993) presented above. Because the viewpoint is moved from one mosaic patch to another, this approach is only useful for flat objects being no larger than the travel capacity of the translation equipment. Furthermore, compensations for lens characteristics such as vignetting which affects image quality near the borders of each patch should be applied. However, there is no theoretical limit on image resolution using this technique.

- Both the micro- and the macro-scanning take place behind the lens. This is the method proposed by the authors. This approach allows for imaging 3D objects of arbitrary size. The resolution of the system is, however, now limited by the diffraction and the image field size of the lens. Such a camera is being developed by the authors, attaining a resolution of $20k \times 20k$ pixels, and a full-size scanning time of about 5 minutes.

They mention that a conversion from RGB to CIELAB will be performed on-line, but give no details on the colorimetric calibration necessary for this conversion.

Lenz *et al.* (1996b) describe the calibration and characterization of this camera, applied on the production of an art paintings catalogue. Two color charts were used for the characterization, the Macbeth ColorChecker with 24 patches, and a MARC chart with 112 colors specifically designed to contain colors used in paintings. Two different illuminations were used, HMI light[3] and 3200K tungsten light.

To determine an analytical mapping from raw camera data (RGB) to XYZ, various variants of first, second and third order transforms were investigated.

The proposed method seems to consist of two steps, *i)* a characterization performed once using the MARC chart and a full third order transform, and *ii)* a simpler calibration, called *matrix white balance*, using the Macbeth chart imaged beside each painting. This approach is preferred over an approach imaging the MARC color chart beside each painting, and performing a full third order correction for each image because of specular reflection on the MARC chart under otherwise optimal lamp positions.

Their best results are RMS $\Delta E_{ab}^* = 3.1$, measured on all colors of the MARC chart taken under HMI illuminations. The result using tungsten light is considerably worse.

4.3   Experimental setup and initial calibration

The digital camera we used is a Kodak Eikonix 1412 camera with a Nikon lens (50mm 1/2.8). It is equipped with three built-in RGB filters mounted on a wheel and with a linear CCD array of 4096 light sensible square elements. A stepper motor moves the array perpendicularly, scanning the image plane in 4096 lines. The analogical signal of each element is AD converted into 12 bit and corrected by the camera hardware with a linear transformation according to a dark current offset and a gain factor which

---

[3]The HMI metallogen lamp developed by Osram Corporation, http://www.osram.com, has a color temperature of approximately 5600K. In the name HMI, the H is an abbreviation for mercury (Hg), M indicates the presence of metals, and I refers to the addition of halogen components such as iodide.

can be numerically adjusted for each individual element. The camera is connected to a PC and driven by software.

4.3.1 General setup

The painting to be digitized is installed vertically on an easel, preferably without its external frame. An ANSI IT8.7/2 (1993) color target is fixed just above it, in the same plane as the painting surface. The distance of the camera to the painting is chosen according to the desired resolution: values between 5 to 10 pixels per mm are typically used. In order to avoid geometrical distortions, the painting has to be placed perpendicularly to the optical axis of the camera. This can be set precisely when the painting is taken in one view by controlling the distance of its four corners to the center of the camera lens. When the painting needs to be taken in several views, the positioning constraints become too strong for building a mosaiced image directly from a juxtaposition of the recorded subimages. A geometrical correction must be done by software to match the overlapping parts with sub-pixel accuracy.

The illumination of the painting is a rather delicate matter, when using a linear array camera that requires long exposure time. It has to be powerful enough to provide a signal far above the dark current of the CCD, but it should also avoid any risk of deterioration of the painting by satisfying experimental condition constraints (illumination < 600 lux, room temperature $< 25°$C, hygrometry $< 50\%$). In the VASARI project 4 quartz tungsten-halogen lamps of 1 kilowatt were used in indirect lighting. With a color temperature of 3200K, their energy in the blue domain where the CCD sensitivity is low was just sufficient. For the Corot paintings we used an HMI daylight lamp with a color temperature of 5500K. The energy in the 3 channels of the camera was well equilibrated. But due to a technical accident we used only a single lamp with a direction lighting of $45°$ to the painting surface. The spatial repartition of the lighting on the painting surface was very inhomogeneous, but was successfully corrected by a processing described in Section 4.4.1.

4.3.2 CCD calibration

After positioning the camera, the painting and the lighting, the general set-up terminates with the settings of the lens aperture and of the camera integration time successively for each of the three channels. These settings must be chosen in such a way that the integration time is minimal and the image signal is as high as possible but without saturation. The integration time must also be a multiple of 0.01 second to limit any coupling of the lighting with the 50Hz alternating current. The general set-up is resumed in the first row of Figure 4.3.

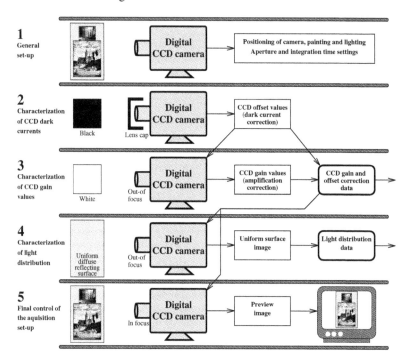

Figure 4.3: *General setup and CCD calibration.*

The following two steps of the CCD calibration are the characterizations of the dark current (offset) and the gain of the analogical signal delivered by the linear CCD array elements. Their values being a function of the integration time and the temperature, we characterize these CCD parameters prior to each acquisition and for each individual element and for the integration time chosen for each one of the three channels. We first statistically estimate the dark current values by measuring the element responses in the black. The resulting values are stored to be re-used as offset correction data (see row 2 of Fig. 4.3). We had previously experimentally verified that the offset corrected signal of an element becomes well-linear with the energy of the light received, *cf.* Section 3.2.2.

We then characterize the electronic gain of each element with the defocused image of a diffuse white chart placed in front of the painting and thus with similar lighting characteristics. We record a set of measurements for a given position of the linear array, and for each element we determine the pixel mean value of its offset corrected values (see row 3 of Fig. 4.3). The resulting curve for the 4096 elements is very jittered due to the variation of the gain from element to element. Fig. 4.4(a) shows a portion of this curve for 200 successive elements (from column 100 to 300). We smoothen this curve by a spatial lowpass filter, in order to retrieve a smooth curve that is a satisfactory approximation of the unknown spatial energy distribution of the defocused white image area covering the current position of the linear array. A data set of 4096 gain correction factors is computed by dividing the filtered curve by the jittered one and is stored with the gain correction data set (see Fig. 4.4(b)).

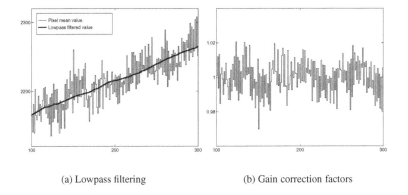

(a) Lowpass filtering   (b) Gain correction factors

Figure 4.4: *Characterization and correction of the CCD electronic gain (simulated data).*

The fourth step is the characterization of the light distribution. We place a large, uniform, light (but not necessarily white), diffuse reflecting board in place of or close in front of the painting. We leave the camera slightly out-of-focus in order to eliminate any local inhomogeneity of the diffusing surface (spot, particle, fiber, etc). We then scan, in the channel delivering the highest signal, the corresponding 12-bit pixel values corrected for offset and gain, and record them in a numerical image denoted LUM. Each pixel value of LUM is proportional to the lighting energy received at the corresponding point on the diffuse surface. LUM and its maximum value constitute the light distribution data (see row 4 of Fig. 4.3). Finally we carefully focus the image of the painting and control the acquisition set-up by verifying that no saturation occurs in any of the three channels (see row 5 of Fig. 4.3).

4.4 Correction algorithms

When the experimental set-up and the CCD calibration have been carried out, we proceed with the scanning of the painting itself. It results in three RGB images, each composed of 12 bit/pixel values already corrected by hardware with CCD gain and offset correction data. These RGB images then undergo a set of three successive corrections, which are described in the next three sections. A flowchart of this process is shown in Fig. 4.5.

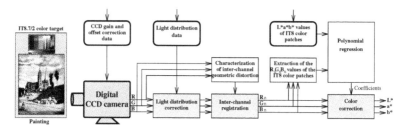

Figure 4.5: *Flowchart of the main data processing. Three successive correction algorithms are applied to the images: light distribution correction, inter-channel registration and color correction.*

4.4.1 Light distribution correction

The first correction eliminates the light distribution inhomogeneities. It is done by dividing each RGB value by the value of the relative light distribution image (LUM normalized by its maximum) at the corresponding pixel. This correction would transform the light image itself in a perfect constant image equal to the maximum of LUM. We can archive the resulting light-corrected RGB images with 3x12 bit quantised values and/or directly continue the following steps of the processing in floating point precision, avoiding thus the introduction of any further quantization effects.

4.4.2 Inter-channel registration

A misalignment of the three channels occur whenever the color planes are not perfectly registered. There are two main causes for this (Khotanzad and Zink, 1996), physical and optical misalignment. The misalignment problem is studied in the field of color segmentation (Khotanzad and Zink, 1996, Hedley and Yan, 1992, Marcu and Abe, 1995)

Physical misalignment occurs if a relative movement between the sensor and the target takes place, typically in a three-pass flatbed scanner, or with a digital multi-pass camera on a not-so-very rigid tripod.

Optical misalignment is due to the prism effect of the lens material (see Figure 4.6). The light rays with different wavelengths are refracted (bent) differently by the lens, and thus hitting the CCD at slightly different locations. This effect is known as "lateral chromatic aberration".

The use for each channel of a specific optical filter in front of or behind the lens (as is the case with the Kodak Eikonix camera) introduces inevitably some chromatic aberrations in the optical path. As a consequence we can observe that the three channels are not perfectly registered, the discrepancies corresponding in particular to tiny differences in the magnification. Radial shifts of about 2 pixels are commonly encountered between two channels on the border of a 4k x 4k image.

To correct these geometrical effects we register the R and B channels on the G one respectively by two polynomial transformations of degree 2 in the image space coordinates. To determine the coefficients of these transformations we first detect with classical picture processing techniques the main characteristic points corresponding to strong local features such as edge corners in each of the 12-bit images (the images corrected for light distribution if available, otherwise the uncorrected scanned ones). For each transformation (R/G and B/G respectively) we then match the best pairs of characteristic points by correlation techniques under the hypothesis of small shifts. We finally use robust statistics and linear RMS techniques for the estimation of the transformation coefficients. The resulting inter-channel geometric registration increases the sharpness of the image and limits the iridescences along the edges between contrasted regions, as shown in Figure 4.7.

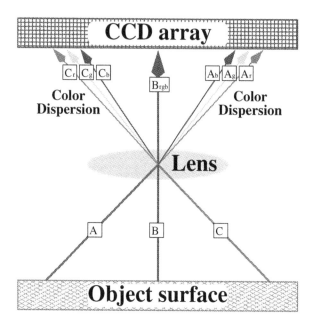

Figure 4.6: *RGB misalignment by optical color dispersion. (Adapted from Khotanzad and Zink, 1996).*

4.4.3   Colorimetric correction

The third correction is a color correction. The analysis of the spectral responses of the digital camera, *cf.* Section 6.2.2, shows that they are far from being linear combinations of the color matching functions of the CIE-XYZ-1931 standard observer.

In order to increase the quality of our color data we characterize the digital camera by using an analytical model, as described in Chapter 3, based on the minimization of the mean square error of a set of measurement points by polynomial regression. By doing so, we achieve a very high colorimetric fidelity, with a mean ΔE_{ab} of approximately 1.5. The exact results vary from image to image.

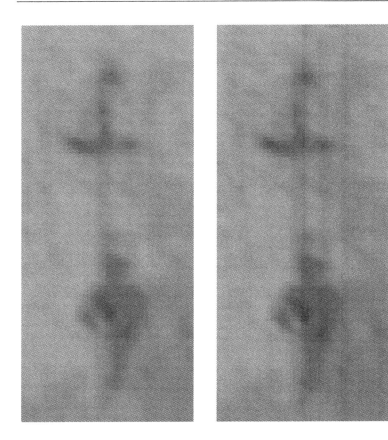

Figure 4.7: *Example of inter-channel registration. Magnified detail of the spear of the Cathedral of Chartres by C. Corot. The original image (left) shows artifacts due to misalignment, e.g. irisation and blurredness. In the empirically corrected image (right) these artifacts are greatly reduced.*

4.5   Post-processing

The CIELAB digital image resulting from the main processing can be archived. It is generally further processed for a given application. Various examples are presented in Fig. 4.8 such as an image mosaicing when very high definition is required, a color conversion when the image has to be visualized on a specific device, or a color facsimile transmission.

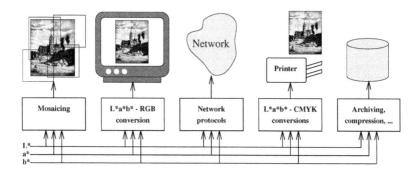

Figure 4.8: *Various post-processing algorithms applied to the acquired CIELAB images.*

4.5.1   Mosaicing

When the dimension of a painting requires more pixels than those provided by the digital camera, we need to perform a mosaicing. This implies making several acquisitions covering the entire surface of the painting. In order to maintain the validity of the CCD calibration data we follow the same general set-up, in particular by keeping the same light distribution on the area of the painting viewed by the camera, and we just translate the painting parallelly to its surface in front of the camera. However it is impossible to guarantee this translation as exactly parallel to the image plane. Small homographic distortions occur between adjacent images and thus geometric corrections have to be made for their registration before building the mosaic. By reserving a large enough overlap between two adjacent images we can match in their overlapping part a set of corresponding feature points and determine a geometric transformation of the second image for it to coincide with the first one. For that we follow a similar approach to the one used for the inter-channel registration. We start by choosing as fixed the central CIELAB image of the painting and we progressively build around the mosaic by repeating the registration step between each remaining CIELAB image and the set of the already registered images.

4.5.2   Visualization and reproduction

We used for the visualization in our experiments a BARCO Reference Calibrator monitor for which we know precisely the phosphor color point coordinates and the specific gamma of each channel. It is calibrated for several white light references, in particular for the daylight D50. For each CIE-LAB triplet we directly derive, from computation and by using an usual CRT model, the corresponding digital RGB gamma-corrected values for the D50 calibrated monitor, *cf.* Section 2.5.3.1. Other transformations of the CIELAB triplets can also be made, such as the conversion to a color printer as explained in Chapter 5.

4.5.3 Colorimetric analysis of fine art paintings

We have proposed a methodology for using a computer to assist in the colorimetric analysis of fine art paintings (Crettez *et al.*, 1996, Crettez, 1998, Hardeberg and Crettez, 1998, Crettez and Hardeberg, 1999). This analysis provides valuable information on the colors in a painting, their distribution, the techniques applied by the artist etc.

We perform a segmentation of the CIELAB space into different regions, such as light and dark colors, pastels and saturated colors etc. This segmentation in CIELAB space also provides a segmentation of the painting itself. We can then perform a colorimetric analysis of the resulting regions separately, and extract several properties, such as the precision of the nuances, color harmonies, principal colors etc. We can also perform statistical analyses of the color distributions.

The colorimetric analysis furthermore allows the demonstration and evaluation of different perceptual effects, such as simultaneous contrast, known from the theories of color appearance.

This methodology has been applied to several paintings, *e.g.* by Francisco de Goya (1746-1828), Jean-Baptiste Camille Corot (1796-1876), and Georges de La Tour (1593-1652). Interesting colorimetric information has been obtained, see Figure 4.9.

4.6 Conclusion

The complete process for the acquisition and the processing of high quality digital color images provides satisfactory results. The described methods have been applied to fine-art paintings on several occasions, for example for the making of a CD-ROM on the French painter Jean-Baptiste Camille Corot (1796-1876) in collaboration with the LRMF (Crettez *et al.*, 1996).

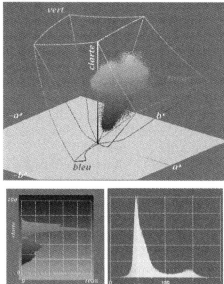

Figure 4.9: *Colorimetric analysis of "Le Forum" by Jean-Baptiste Camille Corot. The color distribution in CIELAB space, the lightness histogram, and the hue angle histogram of the blue/yellow sky (Hardeberg and Crettez, 1998).*

We used in these studies a digital camera with a linear array which requires a long exposure time for the scanning of the painting. An improvement would be to use a digital camera with a large rectangular CCD array up to 4k x 4k, now available.

We would also like to improve the visualization and adapt it to the surrounding conditions of viewing by using advanced color appearance models (see Fairchild, 1997). For the acquisition itself we develop multi-spectral image techniques by increasing the number of filters in order to reconstruct, from the multi-channel values recorded at each pixel, the spectral reflectance curve of the painting at the corresponding point, as described in Chapter 6.

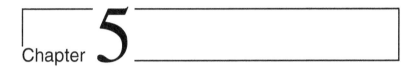

Chapter 5

Colorimetric printer characterization

A NOVEL METHOD FOR THE COLORIMETRIC CHARACTERIZATION OF A PRINTER IS PROPOSED. IT CAN ALSO BE APPLIED TO ANY OTHER TYPE OF DIGITAL IMAGE REPRODUCTION DEVICE. THE METHOD IS BASED ON A COMPUTATIONAL GEOMETRY APPROACH. IT USES A 3D TRIANGULATION TECHNIQUE TO BUILD A TETRAHEDRAL PARTITION OF THE PRINTER COLOR GAMUT VOLUME AND IT GENERATES A SURROUNDING STRUCTURE ENCLOSING THE DEFINITION DOMAIN. THE CHARACTERIZATION PROVIDES THE INVERSE TRANSFORMATION FROM THE DEVICE-INDEPENDENT COLOR SPACE CIELAB TO THE DEVICE-DEPENDENT COLOR SPACE CMY, TAKING INTO ACCOUNT BOTH COLORIMETRIC PROPERTIES OF THE PRINTER, AND COLOR GAMUT MAPPING.

5.1 Introduction

The characterization of a color output device such as a digital color printer defines the relationship between the device color space and a device-independent color space, typically based on CIE colorimetry. This relationship defines a (forward) printer model. Several approaches to printer modeling exist in the literature. They may be divided into two main groups:

■ **Physical models.** Such models are based on knowledge of the physical or chemical behavior of the printing system, and are thus inherently dependent on the technology used (ink jet, dye sublimation, etc.). An important example of physical models for halftone devices is the Neugebauer model (Neugebauer, 1937, Kang, 1997), which treats the printed color as an additive mixture of the tristimulus values of the paper, the primary colors, and any overlap of primary colors. More recent applications of analytical modeling are illustrated with a study of Berns (1993b), who applies a modified version of the Kubelka-Munk spectral model (Kubelka and Munk, 1931) to a dye diffusion thermal transfer printer.

■ **Empirical models.** Such models do not explicitly require knowledge of the physical properties of the printer as they rely only on the measurement of a large number of color samples, used either to optimize a set of linear equations based on regression algorithms, or to build look-up tables for 3D interpolation. Regression models have not been found very successful in printer modeling (Hung, 1993), while the look-up table method is used by several authors, for example Hung (1993) and Balasubramanian (1994).

However, both these groups of printer models have to be inverted to be of practical use for image reproduction, since what we typically need is to transform images colorimetrically defined in a given color space into the color space specific to the printer. The solution to this inverse problem is difficult to find. Iterated optimization algorithms are often needed to determine the device color coordinates which reproduce a given color defined in a device-independent color space, as proposed for example by Mahy and Delabastita (1996).

Another issue that cannot be avoided when discussing printer characterization is gamut mapping. The color gamut of a device such as a printer is defined as the range of colors that can be reproduced with this device. Gamut mapping is needed whenever two imaging devices do not have coincident color gamuts, in particular when a given color in the original document cannot be reproduced with the printer that is used.

We have chosen to use an empirical model to characterize a printer (Hardeberg and Schmitt, 1996; 1997, Schmitt and Hardeberg, 1997; 1998).

The main reason for this is that an empirical model is versatile. It may be applied to printers using different printing technologies, and even to other types of image reproduction devices, such as monitors. In comparison, state-of-the-art physical printer models are limited to one printing technology. Furthermore, the determination of the inverse transformation with physical models requires very extensive computation with non-linear optimization techniques, which we prefer to avoid.

The proposed characterization technique based on an empirical model provides a practical tool to transform colors between any two color spaces, for example between scanner RGB space and printer CMY. In a color management application, it is preferred to connect the device-dependent color representations to some device-independent color space (MacDonald, 1993a, Murch, 1993, Hardeberg *et al.*, 1996, Hardeberg and Schmitt, 1998). We have chosen the CIELAB space (CIE 15.2, 1986) for this purpose since it is used extensively both in literature and industry. Thus our characterization technique provides the transformation between any color point in CIELAB space and the corresponding printer CMY values needed to reproduce the given color. This process also includes a color gamut mapping technique, which can be of any type.

We use an approach based on computational geometry with which we construct two 3D structures, which cover both the entire definition domain of the CIELAB space and the printer color gamut. It provides us with a partition of the space into two sets of non-intersecting tetrahedra, an **inner structure** covering the printer gamut, and a **surrounding structure**, the union of these two structures covering the entire definition domain of the CIELAB space. These 3D structures allow us to easily determine if a CIELAB point is inside or outside the printer color gamut, to apply a gamut mapping technique when necessary, and then to compute by irregular tetrahedral interpolation the corresponding CMY values. We establish thus an empirical inverse printer model.

In the next sections we describe the proposed method, starting by giving an overview of the methodology in Section 5.2. In Section 5.3 and 5.4 we present the construction of the inner structure and the surrounding structure, respectively. We describe in Section 5.5 how we calculate, by tetrahedral interpolation, the transformation from CIELAB to CMY values for any point belonging to the definition domain of CIELAB space, this point being either inside or outside of the color gamut.

5.2 Methodology overview

Our method, as presented in Figure 5.1, consists of first printing a numerical color chart (the input data) covering the entire color gamut of the printer to be characterized. Then we measure colorimetrically the printed chart to obtain the CIELAB values corresponding to each sample. When this is done we dispose, for each color sample of the chart, of their CIELAB values and their corresponding CMY values. Storing these values in a look-up table, we could thus easily establish an empirical forward printer model for converting from CMY to CIELAB, using an interpolation technique (Kanamori *et al.*, 1990, Hung, 1993, Rajala and Kakodkar, 1993, Balasubramanian, 1994, Motomura *et al.*, 1994, Fumoto *et al.*, 1995, Kasson *et al.*, 1995).

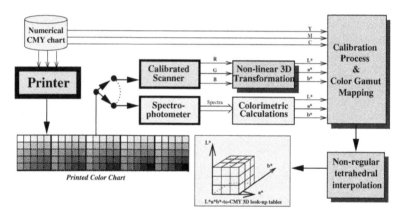

Figure 5.1: *Printer characterization. The numerical color chart is printed, the output is analyzed, and an empirical inverse printer model is established using irregular tetrahedral interpolation.*

However, in practice, we have to convert colorimetric values in the other direction, from CIELAB to CMY. We are then much more interested in establishing directly an empirical inverse printer model.

The main step of the proposed algorithm is the construction of a valid partition of the CIELAB space. A naive approach to this problem would be to apply a 3D Delaunay triangulation directly to the measured CIELAB values. However, this would not suit our purposes, mainly because the gamut is generally not a convex hull in CIELAB space, and then the gamut boundaries would not be correctly represented. In particular, any concavities of the gamut surface would be filled, and the corresponding information about the gamut surface would be lost.

We propose an indirect approach where we apply a 3D Delaunay triangulation in CMY space by taking the CMY triplets from the input data as vertices. Using this 3D triangulation, we would be able to calculate the corresponding CIELAB values for a given CMY triplet simply by barycentric interpolation of the CIELAB values of the vertices of the tetrahedron surrounding the CMY triplet, as was also proposed by Bell and Cowan (1993). This would directly provide us with a forward printer model. But because we are interested in the inverse printer model, we transport the CMY triangulation into CIELAB space by simply replacing the CMY vertices of the triangulation by their measured CIELAB counterparts. This corresponds to a geometric deformation of the triangulation of the gamut cube in which the external boundaries are preserved, as shown in Figure 5.2. The construction of this inner structure will be described in Section 5.3.

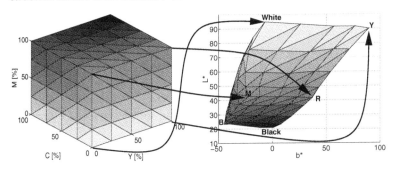

Figure 5.2: *Triangulated CMY color gamut cube (left) and its corresponding geometrically deformed CIELAB color gamut (right).*

At this point we dispose of an inner structure partitioning in tetrahedra the region of the CIELAB space lying inside the printer color gamut. We are able to calculate for any point *inside* the CIELAB color gamut, its corresponding CMY values by tetrahedral interpolation of the CMY values associated with the vertices.

In order to be able to properly treat out-of-gamut colors, we have added a surrounding structure in CIELAB space, defined by a set of six *fictive points* as shown in Figure 5.3. The key issue here is the definition of this surrounding structure in such a way that, together with the inner structure, it defines a valid triangulation which includes the definition domain of the CIELAB space. This will be described in Section 5.4.

5.3 Inner structure

In this section the construction of a valid tetrahedral partition of the printer color gamut in CIELAB space is presented. This corresponds to the inner structure as introduced before.

As seen in the methodology overview, we need first to print an appropriate color chart. This deserves some discussion. Whereas a characterization method using a physical model with few parameters needs only few samples, our method requires that the chart covers the entire color gamut quite densely as it is based on local interpolation. The IT8.7/3 (ANSI IT8.7/3, 1993) input data or parts of it, defined in the CMYK space, could be a good choice for our purpose. However, depending on the precision and the repeatability of the measuring device as well as that of the printer, it may be necessary to remove some colors from the data set. This is motivated by the fact that the characterization process, in particular, the transport of the triangulation from CMY to CIELAB space must remain a topology-preserving *elastic* transformation without any mirroring of a tetrahedron. This will be further elaborated in Section 5.3.2. Furthermore, a too fine subdivision of the CIELAB space would introduce unnecessary complexity to the 3D structures, this being particularly unwanted for the localization algorithm (Section 5.5.1).

A preprocessing of the data set is thus done, by comparing the measured color of all patches of the color chart. If two patches have a color difference less than a threshold ΔE value, the vertex corresponding to one of these patches are removed from the data set. The selection of which color to remove is done by assigning the following priorities to each vertex.

1. **Interior colors.** Colors that belong to the interior of the color gamut, *i.e.* not on the gamut surface.

2. **Face colors.** Colors belonging to one of the six gamut faces, but not belonging to the gamut edges or corners.

3. **Edge colors.** Colors belonging to one of the 12 gamut edges, but not to the gamut corners.

4. **Corner colors.** Colors belonging to one of the 8 gamut corners.

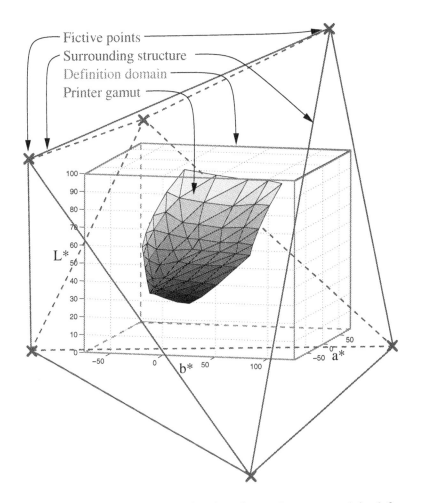

Figure 5.3: *Octahedron surrounding the printer color gamut and the definition domain in CIELAB space. This octahedron, defined by 6 so-called fictive points, together with the triangulation of the volume between its surface and the printer gamut, defines the surrounding structure.*

When two colors are found to be too close, the one with the lowest priority is removed. Remark that the corner colors, with priority 4, will never be removed.

This preprocessing ensures that the data set used for the creation of the data structure described in the following sections, is more coherent, thus limiting possible sources of errors, and avoiding unnecessary data processing.

5.3.1   Delaunay triangulation of the CMY color gamut

To construct our inner structure we use a 3D Delaunay triangulation (Delaunay, 1934, Bern and Eppstein, 1992, Fortune, 1992) in CMY space by taking the CMY triplets from the input data as vertices. The resulting structure is then constituted of a set of tetrahedra (simplices) having the input data as vertices, and whose circumsphere contains no other input point in its interior. This Delaunay property is often denoted as the *empty circumsphere criterion*. Assuming general position of the input CMY points (no five points lie on a single sphere), this defines a unique triangulation. In the degenerate case of co-spherical points any completion of the Delaunay triangulation solves the problem, but the resulting triangulation is no more unique.

We recall that a triangulation can be defined by its adjacency graph whose nodes are the simplices and whose edges are the pairs of adjacent simplices. For a Delaunay triangulation this graph is geometrically realized by the set of the edges and vertices of the associated Voronoï polyhedra. Delaunay (1934) has shown that if for any edge of this graph its two nodes verify the empty circumsphere criterion, then the criterion is verified for any couple of nodes. On this "Lemme général" are based several incremental algorithms (Bowyer, 1981, Watson, 1981, Hermeline, 1982), with which the Delaunay triangulation is built by inserting one by one the 3D points.

For each step of a point insertion, these algorithms are divided in three parts as follows:

1. Localization and deletion of the simplices whose circumsphere contains the inserted point and thus are no more empty. This set of tetrahedra forms a star polyhedron.

2. Creation of new simplices defined by the inserted point and the boundary faces of the star polyhedron, forming a local triangula-

tion of this polyhedron. Determination of the adjacency relations between these new simplices.

3. The union of this local triangulation and the unchanged triangulation of the complementary of the star polyhedron forms the updated Delaunay triangulation. Determination of the adjacency relations between the new simplices and their adjacent unchanged simplices.

We use here the implementation proposed by Borouchaki (1993). To avoid expensive sorting in the second and third parts of the algorithm, the adjacency graph is completed by associating to each simplex a 4×4 matrix which indicates for each one of its four adjacent simplices the indices corresponding to their 3 shared vertices, and to the opposite one (Borouchaki, 1993, Borouchaki *et al.*, 1994).

5.3.2 Transport of the triangulation into CIELAB space

As indicated previously, we transport the CMY triangulation into CIELAB space by simply replacing the CMY vertices of the triangulation by their measured CIELAB counterparts. This corresponds to a geometric deformation of the triangulation of the gamut cube in which the external boundaries are preserved, as shown in Figure 5.2. The resulting triangulation is no more a Delaunay triangulation in CIELAB space, the empty circumsphere criterion being no longer fulfilled. But, it remains generally a valid partition of the CIELAB color gamut, satisfying the following property: the intersection of two of its simplices/tetrahedra is either empty or equal to a vertex, an edge or a face. This implies that no tetrahedron has been mirrored during the transportation from CMY to CIELAB space (see the upper left part of Figure 5.4).

However, in practice, this property must be verified since errors may occur due to either *i)* a too fine subdivision of the gamut, *ii)* measurement errors, or *iii)* strange behavior of either the printer driver software [1] or the physical or chemical properties of the printer itself. For example it has been observed on some color laser printers that, with a specific driver, a regular CMY grid may present a clearly visible luminance order inversion for two particular adjacent patches.

If errors occur, some points of the input data may have to be eliminated from the triangulation. We propose a procedure which checks the validity

---

[1] Sometimes very poor half-toning techniques or a very rough CMY to CMYK conversion are used.

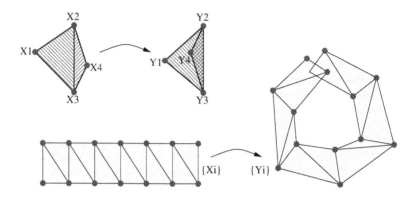

Figure 5.4: *2D illustrations of the no-mirrored-tetrahedron property. Upper left part: After transport, the simplex Y2Y3Y4 is mirrored. The two simplices overlap, thus violating the property of a valid partition. Lower part: A bar-shaped polyhedron {Xi} is transformed into a horseshoe-shaped one {Yi}. Even if, locally, the no-mirrored-tetrahedron property is satisfied, the triangulation is not valid, due to the overlap.*

of the triangulation in CIELAB space by verifying that no tetrahedron has been mirrored. This is denoted the *no-mirrored-tetrahedron property*, and it is verified as follows. For each tetrahedron of the structure we compute the following two determinants:

$$D_{\text{CMY}} = \begin{vmatrix} C_1 & C_2 & C_3 & C_4 \\ M_1 & M_2 & M_3 & M_4 \\ Y_1 & Y_2 & Y_3 & Y_4 \\ 1 & 1 & 1 & 1 \end{vmatrix},$$

$$D_{\text{CIELAB}} = \begin{vmatrix} L_1^* & L_2^* & L_3^* & L_4^* \\ a_1^* & a_2^* & a_3^* & a_4^* \\ b_1^* & b_2^* & b_3^* & b_4^* \\ 1 & 1 & 1 & 1 \end{vmatrix}, \tag{5.1}$$

where (C_i, M_i, Y_i) and (L_i^*, a_i^*, b_i^*), $i = 1 \ldots 4$, are the coordinates of the tetrahedron vertices in CMY space and CIELAB space respectively. If D_{CMY} and D_{CIELAB} have the same sign, the tetrahedron has not been mirrored during the transport. The vertex order chosen to compute the determinants does not matter, as long as it is the same for the two determinants. The absolute value of the determinants is proportional to the volume of the tetrahedron before and after transport respectively (*cf.* Equation 5.21).

It is easy to show that if two adjacent tetrahedra sharing a common face in the CMY triangulation are both not mirrored during the transport, then their intersection in the CIELAB space is strictly equal to the transported common face. If no tetrahedron of the complete 3D structure has been mirrored, then the above property is true for any pair of adjacent tetrahedrons. However, this is not sufficient to guarantee that the complete 3D structure after transport satisfies the property of a valid partition. As an example, let us imagine an elastic distortion of the space, which transforms a bar into a horseshoe as shown in the lower part of Figure 5.4. Locally, the partition property is respected, but if the bending is too strong, the two extremities of the horseshoe may overlap. However, when the convex hull of the initial structure does not auto-intersect during the transport, the no-mirrored-tetrahedron property is sufficient to guarantee that the transported inner structure remains a valid partition. Since the color gamut boundary in CIELAB space generally does not present such degenerate cases, we avoid thus the complete and very expensive checking procedure where any pair of non-adjacent tetrahedrons would be tested for intersection after transport.

However, in some rare cases where the deformation in the CIELAB-space of the color gamut is extremely strong, our approach may fail. We have for example observed that on one particular sublimation color printer some corner colors are strongly desaturated and stand inside the CIE-LAB color gamut. In this particular case, tetrahedra adjacent to such corner points are mirrored. Note that such cases, where the no-mirrored-tetrahedron property is being violated, will be detected by the proposed checking procedure.

5.4   Surrounding structure

The inner structure defines only the triangulation of the color gamut. It is not sufficient to provide a practical tool for gamut mapping, especially for clipping methods. We need efficient means to determine the position in CIELAB space of any color that is located outside the color gamut volume. For this purpose we propose the construction of a *surrounding structure* in CIELAB space which fulfills the two following requirements:

   ▓ It includes completely the current definition domain of the CIELAB-space, which is assumed to include the gamut of any printer (see discussion below).

■ It provides an efficient data structure which allows to navigate easily around and inside the color gamut and to implement any geometrical algorithm for gamut mapping, both continuous and clipping methods.

The definition domain of the CIELAB space depends mainly of the application fields and of the standards that are used in this field. For color facsimile communication services the default gamut range is defined as $L^* \in [0, 100]$, $a^* \in [-85, 85]$, $b^* \in [-75, 125]$, and the components L^*, a^* and b^* are each linearly encoded on 8 or 12 bits (ITU-T T.42, 1994). This basic gamut range is chosen to span the union of available hard copy device gamuts (Beretta, 1996). Thus the color gamut of a given printer can be supposed strictly included in this definition domain.

The surrounding structure is defined by a set of *fictive points* outside of the color gamut as shown in Figure 5.3, and a partition of the space between the inner structure and the convex hull defined by the fictive points. We will describe here an algorithm to define this surrounding structure so that, together with the inner structure, it defines a valid triangulation, which includes the definition domain of the CIELAB space.

It would be natural to add only a minimal number of fictive points: four ones would be sufficient to construct a surrounding tetrahedron. However, considering the symmetry of the color gamut cube in CMY space, we chose as external structure a dual polyhedron of the cube, *i.e* an octahedron defined by 6 vertices, each vertex being associated to a face of the cube.

But the main problem is to construct a triangulation on the full set of points (the fictive ones and those belonging to the printer color gamut) such that:

■ The extended triangulation is valid, satisfying the partition property defined previously.

■ It respects the color gamut hull, *i.e* the external triangular faces of the color gamut (as obtained in Section 5.3.1) belong to the set of the inner faces of the extended triangulation.

Because of the strong deformation of the color gamut in CIELAB space compared to the regular CMY cube (see Figure 5.3 and Appendix D), the determination of these fictive points is not straightforward. We propose a simple and robust method to determine these fictive points as well as the triangulation of the resulting surrounding structure in a way that apply to most printer gamuts.

5.4.1 Construction of the surrounding structure in CIE-LAB space

We choose in CIELAB space a set of 6 fictive points, associated with the 6 faces of the gamut. It can be shown that the position in CIELAB space of each of the six fictive points must lie in the *kernel* of the corresponding distorted cube face, the kernel of a color gamut face being defined as the set of 3D points from where the complete outward surface of the color gamut face is visible. The kernel can then be defined as the intersection of the external half-spaces defined by the tangent planes of each facet of the color gamut face. It is then a convex hull.

We will assume in the following that the kernel of each of the six color gamut faces is neither empty nor closed and thus that it contains at least one infinite point in a specific direction, from where the outward surfaces of all the triangles covering the corresponding color gamut face are visible. We will then place each of the six fictive vertices of the octahedron sufficiently far away from the color gamut to guarantee that the resulting triangulation is valid.

However, the kernel can be a *closed* convex hull, in the case of a very concave color gamut face. It can also be empty in the case of a very convex face: it exists no point from which the outward surface of the face can be entirely seen (see Figure 5.5). Both situations can be only exceptional for CIELAB color gamut boundaries. They would correspond to degenerate cases of very peculiar printer systems. In such exceptional cases our approach cannot be used. To build a surrounding structure adapted to such situations, we could consider the use of a constrained Delaunay triangulation. But this would require the use of a set of Steiner points, and there is no known robust and efficient algorithm to solve this problem (Bern and Eppstein, 1992, Preparata and Shamos, 1985). We have thus not tempted to develop such an algorithm for cases that will maybe never occur. However, our approach allows us to detect such cases if they do appear.

In the next section we describe a discrete approach in computational geometry, which allows us to easily define for each non-closed kernel a *visibility direction* from which, at the infinite point, the outward surface of the corresponding color gamut face is visible. In Section 5.4.3 we deduce

Concave face with a closed kernel

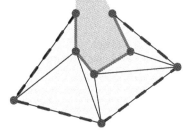

Convex face with no kernel

Figure 5.5: *A 2D example of a structure with two faces. One is very convex, and no kernel exists. The other is very concave, and the kernel is a closed convex hull.*

from these mean directions the position of the fictive points. The construction of the surrounding 3D structure is presented in Section 5.4.4.

5.4.2 Determination of the visibility directions

In the following we will denote F_f, $f = 1 \ldots 6$ the color gamut faces, T_t the tth triangle belonging to the face F_f, and $\mathbf{N}_t$ its outward normal.

Let us consider for each color gamut face F_f a Gauss sphere, *i.e.* a sphere of radius 1 on which are mapped the outwards normal directions of its surface. We use a discrete version of the Gauss sphere by tessellating it in small portions G_p of similar shape and area. Different techniques of tessellation can be used. We have chosen the following one which is simple to implement (Maillot, 1991, Ben Jemaa, 1998): a unit cube parallel to the main axes (L^*, a^* and b^*) is first tessellated following a $\tan(\theta)$ law along its edges, $\theta \in [-\frac{\Pi}{4}, \frac{\Pi}{4}]$, as shown in the left part of Figure 5.6. Then this cube is radially projected onto a co-centered unit sphere. The tangent law provides a regular segmentation of the sphere with quadrangular portions G_p, $p = 1 \ldots 6N_e{}^2$, of nearly constant area, where N_e is the subdivision number chosen for the cube edge ($N_e = 6$ in Figure 5.6). To each portion G_p we associate a flag $\mathcal{F}_p$ and its principal normal direction $\mathbf{N}_p$ defined as the normalized vector sum of the normals associated to its four corners $N_p^{c_i}$, $i = 1 \ldots 4$ (*cf.* Figure 5.7):

$$N_p = \frac{1}{4} \sum_{i=1}^{4} N_p^{c_i} \tag{5.2}$$

This technique allows us to determine quickly (3 tests and 1D look-up table) the portion G_p to which belongs a normal N (Maillot, 1991).

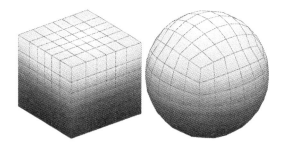

Figure 5.6: *Gauss sphere. A cube tessellated using a tangent law (left) projected onto a unit sphere (right).*

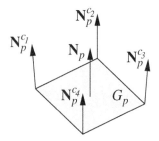

Figure 5.7: *Principal normal direction N_p associated with a Gauss sphere portion G_p.*

The visibility directions from which the infinite point belongs to the kernel of a color gamut face F_f can then be determined as follows. For each portion G_p of the Gauss sphere we first set to TRUE its flag $\mathcal{F}_p$. We then reset it to FALSE as soon as a direction $N \in G_p$ is not located in the positive half-space defined by the outward normal N_t of one of the triangles T_t belonging to the color gamut face F_f (see Figure 5.8).

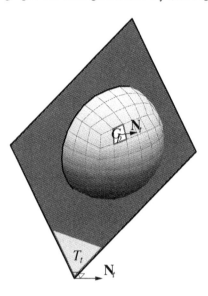

Figure 5.8: *Intersection of the Gauss sphere and the plane defined by the outward normal N_t of the triangle T_t belonging to a face F_f of the printer color gamut. All the normals of the Gauss sphere that are not located in the outward halfspace, correspond to directions from which, at the infinite point, the triangle T_t is not visible. The portions G_p to which belongs such a normal have their flags $\mathcal{F}_p$ reset to* FALSE.

This discrete approach guarantees us that the remaining portions G_p having their flags $\mathcal{F}_p$ still equal to TRUE correspond to sampled directions from which, at the infinite point, the outward surface of F_f can be seen completely. However, due to the tessellation of the Gauss sphere in portion G_p, some valid directions N can be eliminated when a G_p is intersected by the plane defined by the normal N_t of a triangle T_t (see Figure 5.8).

The algorithm is the following:

Algorithm 5.4.1: VISIBILITYDIRECTIONS()

for each face F_f of the color gamut

do $\left\{ \begin{array}{l} \textbf{for each} \text{ portion } G_p \text{ of the Gauss sphere} \\ \text{do} \left\{ \begin{array}{l} \mathcal{F}_p \leftarrow \text{TRUE} \\ \textbf{for each} \text{ oriented triangle } T_t \text{ covering the gamut face } F_f \\ \text{do} \left\{ \begin{array}{l} \textbf{while } \mathcal{F}_p = \text{TRUE} \\ \text{do} \left\{ \begin{array}{l} \textbf{if } \exists \mathbf{N} \in G_p \text{ such that } \mathbf{N} \cdot \mathbf{N}_t \leq 0 \quad\quad\text{(i)} \\ \textbf{then } \mathcal{F}_p \leftarrow \text{FALSE} \end{array} \right. \end{array} \right. \end{array} \right. \end{array} \right.$

The portions G_p being small spherical lozenges, the test given by line (i) of Algorithm 5.4.1 is equivalent[2] to the following, simpler one:

$$\begin{array}{l} \textbf{if } \exists i,\ i \in \{1\dots4\} \text{ such that } \mathbf{N}_p^{c_i} \cdot \mathbf{N}_t \leq 0,\ i = 1\dots4 \\ \textbf{then } \mathcal{F}_p \leftarrow \text{FALSE} \end{array} \quad\quad (5.3)$$

This test can be replaced by another one 4 times quicker but more restrictive (it further eliminates some valid directions $\mathbf{N}$) as follows:

$$\textbf{if } \mathbf{N}_p \cdot \mathbf{N}_t < \sin(\theta) \quad \textbf{then } \mathcal{F}_p \leftarrow \text{FALSE} \quad\quad (5.4)$$

where θ is the maximal angle, among the portions of the Gauss sphere, between the principal direction $\mathbf{N}_p$ and the direction associated to one of its corners:

$$\theta = \max_{\substack{\forall p \\ i=1\dots4}} \text{angle}\left(\mathbf{N}_p, \mathbf{N}_p^{c_i}\right) \quad\quad (5.5)$$

To explain the theoretical basis of this test, let us consider a cone centered on the principal normal $\mathbf{N}_p$ of G_p with an aperture angle of θ. $\mathbf{N} \in G_p$ belongs then surely to this cone (see Figure 5.9). The property

$$\exists \mathbf{N} \in G_p \text{ such that } \mathbf{N} \cdot \mathbf{N}_t \leq 0 \quad\quad (5.6)$$

is then replaced by the larger property

$$\exists \mathbf{N} \in \text{cone}(\mathbf{N}_p, \theta) \text{ such that } \mathbf{N} \cdot \mathbf{N}_t \leq 0. \quad\quad (5.7)$$

This property is equivalent to the following one:

$$\mathbf{N}_p \cdot \mathbf{N}_t \leq \cos(\frac{\pi}{2} - \theta) = \sin\theta \qu\quad (5.8)$$

which justify the test of Equation 5.4.

---

[2]The arc angles of their edges are by construction smaller than $\frac{\pi}{2}$. This guarantees us that each lozenge is enclosed in a half-sphere.

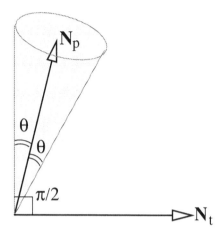

Figure 5.9: *A cone centered on* $\mathbf{N}_p$.

For the construction of the surrounding structure, we have, for each face, to determine on the Gauss sphere, the direction $\mathbf{D}_f$ of an infinite point belonging to its kernel. We could choose any one of the directions belonging to the remaining portions with $\mathcal{F}_p = \text{TRUE}$. However, to guarantee a robust implementation we use the property that a kernel is a convex hull and choose as visibility direction $\mathbf{D}_f$ of the kernel of the cube face F_f the normalized vector sum of the principal normal directions $\mathbf{N}_p$ of the remaining portions G_p with $\mathcal{F}_p = \text{TRUE}$, as indicated in Equation 5.9 and illustrated in Figure 5.10.

$$\mathbf{D}_f = \operatorname*{mean}_{p|\mathcal{F}_p=\text{TRUE}} (\mathbf{N}_p) \qquad (5.9)$$

Because of our discrete approach it could happen that all the portions have been flagged to FALSE although the actual kernel is not a strictly closed hull. This would correspond to a too coarse segmentation of the Gauss sphere. To avoid this problem we just need to segment the Gauss sphere in finer portions. There is then a compromise: to choose between the total number of portions to be used, and the computing time that is proportional to this number. In practice a linear subdivision of the cube edge into 6 elements has been sufficient in all the tested examples (a total of 216 portions).

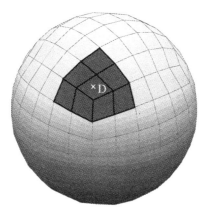

Figure 5.10: *The visibility direction* $\mathbf{D}_f$ *corresponding to the arithmetic mean of the normal vectors* $\mathbf{N}_p$ *that have* $\mathcal{F}_p = \text{TRUE}$.

5.4.3 Determination of the fictive points in CIELAB space

The visibility directions $\mathbf{D}_f$, $f = 1 \ldots 6$ being determined, we have now to define six fictive points satisfying our requirements (situated at finite distance, enclosing the CIELAB definition domain, and belonging to the kernels).

Let us denote $\mathbf{G}$ the CIELAB point of coordinates $(50, 0, 0)$ that corresponds to a medium gray of lightness 50. For each color gamut face $\mathbf{F}_f$ we consider the half-line with $\mathbf{G}$ as zero point and $\mathbf{D}_f$ as direction. Then we determine on this half-line the point $\mathbf{P}_f$ nearest to $\mathbf{G}$ and belonging to the kernel of the color gamut face F_f. This point does exist because the kernel has an infinite point in this direction.

To determine the point $\mathbf{P}_f$ we can consider the following parametric representation of the half line: $\mathbf{P}_f = \mathbf{G} + k_f \mathbf{D}_f$, $k_f \geq 0$. The kernel of a face F_f being the intersection of the kernels of all the triangles T_t covering F_f,[3] we proceed as follows to determine the value of k_f defining $\mathbf{P}_f$. Let $\mathbf{N}_t$ be the outward normal of a triangle T_t and d_t the signed distance between its plane and the origin. A point $\mathbf{P}$ of its plane satisfies then the following equation:

$$\mathbf{P} \cdot \mathbf{N}_t + d_t = 0. \tag{5.10}$$

---

[3] These kernels are the positive half-spaces as illustrated in Figure 5.8.

We first determine the point $\mathbf{P}_t = \mathbf{G} + k_t \mathbf{D}_f$ on the half-line being nearest to $\mathbf{G}$ and belonging to the kernel of the triangle T_t. $\mathbf{P}_t$ being located at the intersection of the half-line with the plane of the triangle, we have the following equation:

$$(\mathbf{G} + k_t \mathbf{D}_f) \cdot \mathbf{N}_t + d_t = 0, \qquad (5.11)$$

from which we deduce its k_t value:

$$k_t = -\frac{\mathbf{G} \cdot \mathbf{N}_t + d_t}{\mathbf{D}_f \cdot \mathbf{N}_t}. \qquad (5.12)$$

The point $\mathbf{P}_f$ with

$$k_f = \max_{T_t \in F_f} (k_t) \qquad (5.13)$$

is then the nearest point on the half-line belonging to the kernel of the color gamut face F_f, as shown (in 2D) in Figure 5.11.

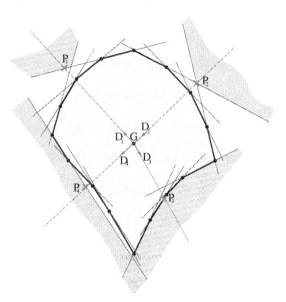

Figure 5.11: *Determination for each gamut faces F_f, of the point $\mathbf{P}_f$ nearest to $\mathbf{G}$, situated on the half-line of direction $\mathbf{D}_f$ and belonging to the kernel of the face F_f (illustration in 2D).*

The six points $\mathbf{P}_f$ constructed as above for the six color gamut faces F_f, $f = 1 \ldots 6$, define an initial octahedron. This octahedron is the dual of the CMY cube, its 8 faces being in duality with the 8 corner points of the color gamut. We will assume in the following that this octahedron encloses the point $\mathbf{G}$. We have never encountered a degenerate case for which this was not the case. But it remains theoretically possible that this could happen in the case of a color gamut strongly distorted in CIELAB space. We do not propose a solution for such an exceptional situation and we just verify this assumption before pursuing the processing.

The problem now is to increase the size of the octahedron such that it encloses the CIELAB definition domain, under the constraint that its six vertices remain in the kernel of the face F_f. According to our assumption on $\mathbf{G}$ it is possible to increase the size of the initial octahedron according a similarity of center $\mathbf{G}$ and coefficient s, $s \geq 1$. With this similarity we guarantee by construction that each of the six new vertices will remain in the kernel of the corresponding color gamut face.

The factor $s \geq 1$ of the similarity must be chosen large enough to allow the transformed octahedron to enclose the definition domain. Various approaches can be chosen. We propose the following (see Figure 5.12).

Let us denote $\mathbf{V}_f$ the new vertices obtained by similarity:

$$\mathbf{V}_f = \mathbf{G} + s(\mathbf{P}_f - \mathbf{G}). \tag{5.14}$$

By construction each point $\mathbf{V}_f$ is on the half-lines of direction D_f associated to the color gamut face F_f, s being greater than 1 and the kernel being a convex hull, $\mathbf{V}_f$ still belongs to its kernel. Considering a translation of the space by the vector $-\mathbf{G}$ and denoting the new points with a prime, we have simply:

$$\mathbf{V}'_f = s\mathbf{P}'_f. \tag{5.15}$$

Let us now consider a face $\mathbf{P}_{f_1}, \mathbf{P}_{f_2}, \mathbf{P}_{f_3}$ of the initial octahedron. We calculate its outward unit normal $\mathbf{N}_{f_1 f_2 f_3}$ and determine the signed distance $d_{f_1 f_2 f_3}$ between its oriented plane and $\mathbf{G}$,

$$d_{f_1 f_2 f_3} = \mathbf{P}'_{f_1} \cdot \mathbf{N}_{f_1 f_2 f_3}. \tag{5.16}$$

Our assumption on $\mathbf{G}$ will be satisfied, *i.e.* $\mathbf{G}$ is inside the initial octahedron, if the 8 signed distances $d_{f_1 f_2 f_3}$ corresponding to the 8 faces of this octahedron are all negative ($d_{f_1 f_2 f_3} \leq 0$). If this condition is fulfilled, we determine their minimum absolute value which is equal to the radius $r_{\mathbf{G}}$ of

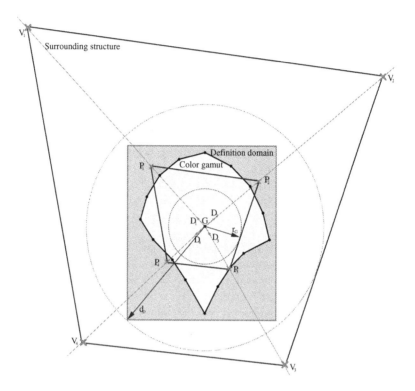

Figure 5.12: *Determination of the final surrounding structure by a similarity operation. See also Figure 5.3 for an illustration of the surrounding structure in 3D.*

the maximal sphere included in the octahedron and of center $\mathbf{G}$:

$$r_{\mathbf{G}} = \min_{f_1 f_2 f_3} |d_{f_1 f_2 f_3}|. \tag{5.17}$$

If this radius is smaller than the maximum distance d_D between $\mathbf{G}$ and the corners of the CIELAB definition domain (as it is usually the case), we choose for the similarity the following coefficient s:

$$s = (1 + \epsilon') \frac{d_D}{r_{\mathbf{G}}}. \tag{5.18}$$

where ϵ' is a small positive value, in order to avoid any computational problems for data near the limits of the definition domain.

This similarity will increase the size of the maximal enclosed sphere of center **G** such that the definition domain will be enclosed by the transformed sphere and thus, *a fortiori*, by the final transformed octahedron.

5.4.4 Triangulation of the surrounding structure

The fictive points being well defined in CIELAB space, we have now to construct the surrounding structure. This could be done directly in CIELAB space according to the inner structure which already has been constructed and which has been used for the determination of the six CIELAB fictive points. For that we would first have to connect each fictive point P_f to the triangles T_t belonging to the corresponding visible face F_f and then fill the remaining concavities with tetrahedra sharing 2 or 3 fictive points (see a 2D illustration in Figure 5.13). This would be theoretically possible but very tedious to implement properly. It is indeed possible to obtain exactly the same surrounding structure by a much simpler construction in CMY space as explained now.

Assume that we have triangulated the cube as described previously. We then first define another set of 6 fictive points in CMY space as indicated in Figure 5.14. We then triangulate the joint set of fictive points and input data by constructing three distinct classes of external tetrahedra, having 1, 2, or 3 vertices being fictive points, and the other vertices being color points belonging to the surface of the color gamut cube, as shown in Figure 5.15:

- The 6 subsets of tetrahedra above each cube face. These tetrahedra have a single octahedron vertex and their opposite triangular face belongs to a face of the cube.

- The 12 subsets of tetrahedra associated to each edge of the cube. These tetrahedra have two octahedron vertices and the two other ones belongs to an edge of the cube

- The 8 remaining tetrahedra associated to each vertex of the cube. These large tetrahedra share 3 vertices with the octahedron and the remaining vertex with the cube.

The fictive octahedron must be big enough to contain the cube. The radius of its circumsphere must thus be $\sqrt{3}$ times larger than the radius of the circumsphere of the cube. In practice we use a factor $2\sqrt{3}$ in order to avoid any numerical problem in the construction of the external tetrahedra

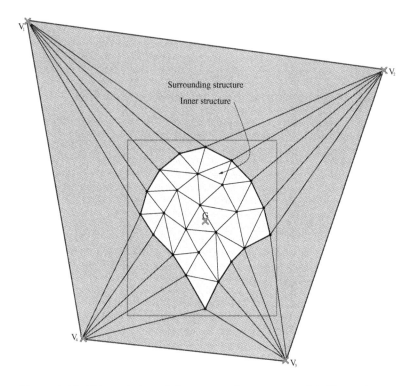

Figure 5.13: *Triangulation of the surrounding structure (2D illustration).*

(see below). The radius of the circumsphere of the octahedron is then equal to 2 times the length of the cube edge (see Figure 5.14.)

We then transport the resulting triangulation to CIELAB space by replacing the CMY vertices (fictive points and input data) with their CIE-LAB counterparts, as described in Section 5.2 and 5.3.2. We thus define a valid triangulation of the joint inner and surrounding structures in CIELAB space.

Different approaches can be chosen for the construction of the surrounding triangulation. A direct construction could easily be made when the color patches on the surface of the CMY color gamut are regularly distributed. But because the cube and the octahedron are both convex hulls, the 3D CMY surrounding triangulation is also a Delaunay triangulation, the criterion of the empty circumsphere being satisfied. It is then more

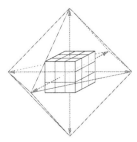

Figure 5.14: *Construction of a surrounding octahedron in CMY space.*

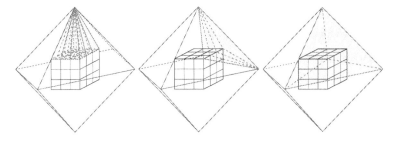

Figure 5.15: *Surrounding structure in CMY space. The 3 classes of external tetrahedra are indicated, having respectively 1, 2, and 3 vertices being fictive points, and the rest belonging to the color gamut*

straightforward to use a Delaunay algorithm to build directly and simultaneously in CMY space the inner and surrounding triangulations on the complete set of points: the color points belonging to the CMY color gamut and the 6 octahedron vertices. We recall that the Delaunay triangulation also has the advantage to work on any irregular distribution of the color points allowing us to add more colors, such as skin tones inside the CMY cube or saturated colors on its surface.

Finally we replace in the 3D data structures the CMY coordinates of the vertices by their CIELAB counterparts, as already described in Section 5.3.2.

5.5   CIELAB-to-CMY transformation

Using these two structures, the inner structure and the surrounding structure as defined above, we are now able to calculate, by tetrahedral interpolation, the transformation from CIELAB to CMY for any point belonging to the definition domain of CIELAB space. This is typically done either

▓ directly for all pixels of an image to be printed, or

▓ for all the vertices of a regular grid composing a CIELAB-to-CMY 3D look-up table (see Appendix B) which can be stored in a device profile and then further used by a color management system (see Section 2.5.1 and *e.g.* MacDonald, 1993a, Hardeberg *et al.*, 1996, ICC.1:1998.9, 1998).

The transformation comprises two main steps, the localization of the tetrahedron enclosing the CIELAB point that is to be transformed (Section 5.5.1), and the irregular tetrahedral interpolation using the vertices of this tetrahedron (Section 5.5.2). Any gamut mapping method may be integrated in this process, as described in Section 5.5.3.

5.5.1   Localization of a CIELAB point in the 3D structure

It would be too expensive to locate the tetrahedron T^P which encloses the CIELAB input point P by checking systematically all tetrahedra T_i. We have therefore implemented the so-called *walking algorithm* as described in the following:

Algorithm 5.5.1: WALKINGALGORITHM()

Pick an initial tetrahedron T_0.[4]
repeat
⎧ **for each** face of the current tetrahedron T_i
⎪   ⎧ **if** P in outward halfspace delimited by the plane of the face
⎨ **do** ⎨ **then** ⎧ T_{i+1} ← neighboring tetrahedron sharing this face
⎪   ⎩   ⎩ Reiterate
⎪ **if** the four tests are negative
⎩   **then** P is located inside the tetrahedron, and T^P ← T_i.
until the enclosing tetrahedron T^P is found[5]
if T_P does not belong to the color gamut[6]
    then P is an out-of-gamut point. Apply a gamut clipping technique.[7]
Calculate output CMY values by interpolation cf. Section 5.5.2.

5.5.2 Irregular tetrahedral interpolation

When the tetrahedron $\mathbf{P}_0\mathbf{P}_1\mathbf{P}_2\mathbf{P}_3$ that belongs to the color gamut and that encloses the point $\mathbf{P}$ (or $\mathbf{P}'$ when $\mathbf{P}$ is out of gamut) is found, the resulting CMY values are calculated by barycentric interpolation as follows: $\mathbf{P}$ (or $\mathbf{P}'$) divides the enclosing tetrahedron into 4 sub-tetrahedrons, as shown in Figure 5.16, each having a volume Δ_i determined by the following determinant,

$$\Delta_i = \frac{1}{6} \begin{vmatrix} \mathbf{P}_i & \mathbf{P}_{i+1} & \mathbf{P}_{i+2} & \mathbf{P} \\ 1 & 1 & 1 & 1 \end{vmatrix}, \quad i = 0\dots3, \tag{5.19}$$

where the indices are taken modulo 3. The positive barycentric coefficients W_i of the interior point $\mathbf{P}$ are then defined by[8]

$$W_i = |\Delta_i/\Delta|, \tag{5.20}$$

where Δ is the volume of the tetrahedron $\mathbf{P}_0\mathbf{P}_1\mathbf{P}_2\mathbf{P}_3$, given by

$$\Delta = \frac{1}{6} \begin{vmatrix} \mathbf{P}_0 & \mathbf{P}_1 & \mathbf{P}_2 & \mathbf{P}_3 \\ 1 & 1 & 1 & 1 \end{vmatrix}. \tag{5.21}$$

The final output values C, M, and Y are then calculated as follows:

$$C = \sum_{i=0}^{3} W_i C_{\mathbf{P}_i}, \; M = \sum_{i=0}^{3} W_i M_{\mathbf{P}_i}, \; Y = \sum_{i=0}^{3} W_i Y_{\mathbf{P}_i}, \tag{5.22}$$

where $C_{\mathbf{P}_i}$, $M_{\mathbf{P}_i}$, and $Y_{\mathbf{P}_i}$ are the CMY values associated with the tetrahedron vertices $\mathbf{P}_i, i = 0\dots3$.

---

[7]If the algorithm is applied for the transformation of an image, T_0 should be chosen as the tetrahedron enclosing the previous pixel. Statistically the new color will be near the previous one.

[7]By construction the enclosing tetrahedron exists when $\mathbf{P}$ belongs to the CIELAB definition domain.

[7]A flag in the 3D data structure informs if a tetrahedron belongs to the inner triangulation or not.

[7]Either the mapping is defined explicitly as $\mathbf{P}' = f(\mathbf{P})$. In this case we reiterate the whole algorithm with $\mathbf{P}'$. Or the mapping is defined geometrically, but necessitate a search, as for example the projection on the nearest point of the color gamut with or without constraints (for example hue, lightness and/or saturation constant). We can then utilize the triangulated structures to implement the gamut mapping algorithm, as discussed in Section 5.5.3.

[8]The absolute values are used because the volumes Δ_i are signed according to the order of the vertices of the determinant (Eq. 5.19).

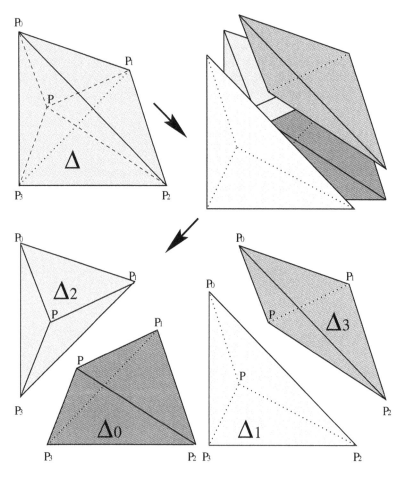

Figure 5.16: *The division of a tetrahedron* $\mathbf{P}_0\mathbf{P}_1\mathbf{P}_2\mathbf{P}_3$ *into four sub-tetrahedra* $\mathbf{P}\mathbf{P}_1\mathbf{P}_2\mathbf{P}_3$, $\mathbf{P}_0\mathbf{P}\mathbf{P}_2\mathbf{P}_3$, $\mathbf{P}_0\mathbf{P}_1\mathbf{P}\mathbf{P}_3$ *and* $\mathbf{P}_0\mathbf{P}_1\mathbf{P}_2\mathbf{P}$, *defined by the input point* $\mathbf{P}$.

5.5.3   Color gamut mapping

Gamut mapping is needed whenever two imaging devices do not have coincident color gamuts, in particular when a given color in the original document cannot be reproduced with the printer that is used. Several researchers have addressed this problem, see for example the following references: (Stone *et al.*, 1988, Gentile *et al.*, 1990, Stone and Wallace, 1991, Pariser, 1991, Hoshino and Berns, 1993, MacDonald, 1993b, Wolski *et al.*, 1994, Spaulding *et al.*, 1995, MacDonald and Morovič, 1995, Katoh and Ito, 1996, Luo and Morovič, 1996, Montag and Fairchild, 1997, Tsumura *et al.*, 1997, Morovic and Luo, 1997; 1998).

Gamut mapping techniques may be divided into two categories, although an efficient practical solution is likely to be a combination of these two categories:

■ **continuous methods** applied to all the colors of an image, such as gamut compression and white point adaptation, and

■ **clipping methods**, applied only to colors that are out of gamut.

A brief presentation of the various techniques of color gamut mapping is given in Appendix E.

The 3D structures allow us to implement easily any gamut mapping technique, such as those mentioned above. Our geometrical approach is particularly well adapted to a combination of continuous and clipping methods.

When needed, a continuous gamut mapping technique may be applied to each input point prior to the interpolation described above. However, if the inverse gamut mapping function exists, it is more computationally effective to apply it to the CIELAB vertices of the 3D structures, as shown for the case of a simple compression in Figure 5.17. The advantage of this approach is to directly include the gamut mapping transformation into the localization step (Section 5.5.1). However, here also it is important to verify that no tetrahedron is mirrored during the inverse gamut mapping transformation in order to preserve the validity triangulation in CIELAB space. Other continuous gamut mapping techniques such as 3D morphing (Spaulding *et al.*, 1995) can also be applied.

If, after the continuous gamut mapping, the input color point is still out of gamut, *i.e.* T^P belongs to the surrounding structure, as already discussed in the previous section, a gamut clipping method must be applied. For example a radial clipping (Pariser, 1991) is easily effectuated by 'walking' from tetrahedron to tetrahedron in the surrounding structure, following a line from **P** towards a mid-gamut point until we reach the color gamut boundary by encountering a tetrahedron belonging to the inner structure.

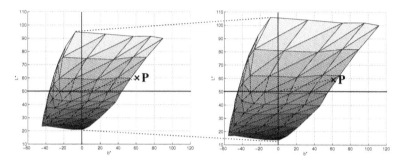

Figure 5.17: *A gamut compression of 20% performed by applying the inverse compression onto all the vertices of the 3D structures (only the inner structure is represented here).*

5.6 Conclusion

The proposed printer characterization method presents several strong points of interest. First, it performs efficiently the inverse transformation from CIELAB (or any other 3D color space) to CMY directly without using numerical optimization techniques. Secondly it is able to easily incorporate different gamut mapping techniques, both continuous and clipping methods. Thirdly it is versatile, not being limited to one specific printing technology. The extension to fourcolour CMYK printers is straightforward when the amount of black ink is determined directly from the CMY values as in the gray-component replacement (GCR) technique (Kang, 1997).

Unfortunately, we have not yet obtained quantifiable experimental results by using this technique. The reason for this is the printers available in our laboratory:

■ An Epson Stylus Color II ink jet printer. The driver software does not permit us to control the amount of inks; an RGB-to-CMYK transformation is always performed. The shape of the CIELAB color gamut thus is not well-behaved, see Appendix D.

▨ A Mitsubishi S340-10 sublimation printer having a nice color gamut (see Appendix D,) offering full control over the ink percentages, but being now out of order.

▨ A Hewlett Packard Color Laser Jet 5M. It is impossible to control directly the amount of ink printed on the paper, some hardware processing is directly done in the printer head, and cannot be bypassed. This results for example in that a regular CMY grid presents a clearly visible luminance order inversion for two particular adjacent green patches. Simple postscript commands that are wrongly executed have shown us that explicitly. We specified one patch with ink densities of $[C, M, Y, K] = [100\%, 0\%, 100\%, 0\%]$, and another with $[C, M, Y, K] = [100\%, 25\%, 100\%, 0\%]$. The second patch should logically be darker than the first, however it turns out to be lighter. This results clearly in the mirroring of a tetrahedron.

Nevertheless, the method is exploited industrially; the developed software has been transferred to the company Couleur, Communication, Ecriture (CCE) S.A.R.L., which has included it in their commercial color facsimile and color management software. Some foreign companies have also showed their interest. A European patent is pending on the procedure (Schmitt and Hardeberg, 1997; 1998).

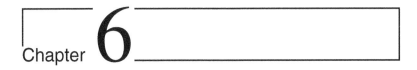

Chapter **6**

Multispectral image acquisition: Theory and simulations

IN THIS CHAPTER WE DESCRIBE A SYSTEM FOR THE ACQUISITION OF
MULTISPECTRAL IMAGES USING A CCD CAMERA WITH A SET OF OP-
TICAL FILTERS. A SPECTRAL MODEL OF THE ACQUISITION SYSTEM
IS ESTABLISHED, AND WE PROPOSE METHODS TO ESTIMATE ITS SPEC-
TRAL SENSITIVITIES BY CAPTURING A CAREFULLY SELECTED SET OF
SAMPLES OF KNOWN SPECTRAL REFLECTANCE AND BY INVERTING
THE RESULTING SYSTEM OF LINEAR EQUATIONS. BY SIMULATIONS
WE EVALUATE THE INFLUENCE OF ACQUISITION NOISE ON THIS PRO-
CESS. WE FURTHER DISCUSS HOW THE SURFACE SPECTRAL REFLECT-
ANCE OF THE IMAGED OBJECTS MAY BE RECONSTRUCTED FROM THE
CAMERA RESPONSES. WE PERFORM A THOROUGH STATISTICAL ANA-
LYSIS OF DIFFERENT DATABASES OF SPECTRAL REFLECTANCES, AND
WE USE THE OBTAINED STATISTICAL INFORMATION TO CHOOSE A SET
OF OPTIMAL OPTICAL FILTERS FOR THE ACQUISITION OF OBJECTS OF
THE SAME TYPE AS THOSE USED FOR THE REFLECTANCE DATABASES.
FINALLY WE PRESENT AN APPLICATION WHERE THE ACQUIRED MUL-
TISPECTRAL IMAGES ARE USED TO PREDICT CHANGES IN COLOR DUE

TO CHANGES IN THE VIEWING ILLUMINANT. THIS METHOD OF ILLU-
MINANT SIMULATION IS FOUND TO BE VERY ACCURATE, AND WORK-
ING ON A WIDE RANGE OF ILLUMINANTS HAVING VERY DIFFERENT
SPECTRAL PROPERTIES.

6.1　Introduction

A multispectral image is an image where each pixel contains information
about the spectral reflectance of the imaged scene. Multispectral images
carry information about a number of spectral bands: from three components per pixel for RGB color images to several hundreds of bands for hyperspectral images. Multispectral imaging is relevant to several domains
of application, such as remote sensing (Swain and Davis, 1978), astron-
omy (Rosselet et al., 1995), medical imaging, analysis of museological
objects (Maître et al., 1996, Haneishi et al., 1997), cosmetics, medicine
(Farkas et al., 1996), high-accuracy color printing (Berns, 1998, Berns
et al., 1998), or computer graphics (Peercy, 1993). Hyperspectral image
acquisition systems are complex and expensive, limiting their current use
mainly to remote sensing applications. Multispectral scanners are mostly
based on a point-scan scheme (Harding, 1997, Manabe et al., 1994, Manabe and Inokuchi, 1997), and are thus too slow for our applications.

Rather than such point-scan systems, we used an approach in which
a set of chromatic filters is used with a CCD camera. It is well known
that with 3 well-chosen filters, it is possible to obtain a fairly good reconstruction of the color tristimulus values of the reference human observer as
defined in colorimetry. Rather than just reconstructing colorimetric tristimulus values, our aim is to estimate the *spectral reflectance* curve by using
more than three filters in sequence. We propose a solution where the filters
are chosen sequentially from a set of readily available filters. This choice is
optimized, taking into account the statistical properties of the object spectra, the spectral characteristics of the camera, and the spectral radiance of
the lighting used for the acquisition.

The multispectral image acquisition system we describe here is inherently device independent, in that we seek to record data representing the
spectral reflectance of the surface imaged in each pixel of the scene, by
discounting the spectral characteristics of the acquisition system and of

the illuminant. We will suppose that the spectral radiance of the illuminant used for the acquisition is known, either by direct measurement, or indirectly by estimation of the spectral characteristics of the acquisition system. We thus do not discuss here issues such as *computational color constancy* (Maloney and Wandell, 1986, Hurlbert *et al.*, 1994, Funt, 1995, Tominaga, 1996; 1997, Finlayson *et al.*, 1997, Zaidi, 1998), that is, the automatic determination of the color image of a scene as it would have been seen under a standard illuminant, from the camera responses obtained under an arbitrary unknown illuminant.

To obtain device independent image data of high quality, it is important to know the spectral characteristics of the system components involved in the image acquisition process. We propose in the next section an approach to the estimation of the spectral characteristics of the acquisition system, followed in Section 6.3 by a discussion on how the spectral reflectances of actual surfaces may be estimated from the camera responses. In Section 6.4 we perform a statistical analysis of different sets of spectral reflectances. This is an important prerequisite when a multispectral image acquisition system is to be designed, in particular for the choice of the number of image channels to be used. In Section 6.5, we discuss how to choose an optimal set of color filters to be used with a given camera. In Section 6.6, we perform an evaluation of the quality of the entire multispectral image acquisition system, followed in Section 6.7 by an application where the acquired multispectral images are used to simulate the image of a scene, as it would have appeared under a given illuminant.

6.2 Spectral characterization of the image acquisition system

In order to properly calibrate an electronic camera for multispectral applications, the spectral sensitivity of the camera and the spectral radiance of the illuminant should be known.

Theoretically, in the absence of noise, the spectral sensitivities may be obtained to the desired spectral accuracy by measuring a set of P samples of known linearly independent spectral reflectances illuminated by light of known spectral radiance, and by inverting the resulting system of linear equations, for example by using the Moore-Penrose pseudoinverse. Thus the spectral characteristics can be determined up to a sampling rate of N

wavelengths, $N \leq P$. However, in the presence of noise, and due to linear dependence between the sample spectra, computing the pseudoinverse becomes hazardous. We describe several approaches to this problem. In particular we show that the choice of samples is of great importance for the quality of the characterization, and we present an algorithm for the choice of a reduced number of samples.

A common solution for this kind of inverse problem is to use a method based on a singular value decomposition (SVD) and to use only those components whose singular values are greater than a certain threshold value. This solution is often referred to as the principal eigenvector (PE) solution (Farrell and Wandell, 1993, Sharma and Trussell, 1993; 1996c). Unfortunately, by using such a method, the resulting principal vectors are linear combinations of the full set of sample spectra, *i.e.* it is not obvious which of the sample spectra are really relevant to the characterization. This approach requires then the use of a large sample set to provide good results.

We present here an alternative method in which we start from a small number of reflectance samples that are chosen according to their spectral variance. Additional reflectance samples are added in order to maximize the volume defined by the vector space spanned by these spectra. This method not only allows for an efficient estimation of the spectral sensitivity functions of electronic cameras, but also provides a tool for the design of optimized color targets containing only those patches that are most significant for the characterization process.

In order to perform a spectral characterization of an electronic camera, it is necessary to establish a model of the system. In Section 6.2.1 we describe a general spectral model of an image acquisition system. This model applies to several types of image acquisition systems, in particular to electronic CCD cameras. Then, several methods for the spectral characterization are proposed and evaluated in Section 6.2.2, followed by a discussion of the results in Section 6.2.3.

6.2.1   Image acquisition system model

The main components involved in the image acquisition process are depicted in Figure 6.1. We denote the spectral radiance of the illuminant by $l_R(\lambda)$, the spectral reflectance of the object surface imaged in a pixel by $r(\lambda)$, the spectral transmittance of the optical systems in front of the detector array by $o(\lambda)$, the spectral transmittance of the kth optical color filter by $\phi_k(\lambda)$ and the spectral sensitivity of the CCD array by $a(\lambda)$. Note that only one optical color filter is represented in the figure. In a multichannel system, a set of filters is used.

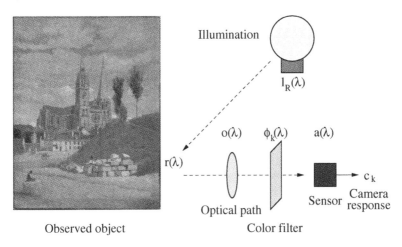

Figure 6.1: *Schematic view of the image acquisition process. The camera response depends on the spectral sensitivity of the sensor, the spectral transmittance of the color filter and optical path, the spectral reflectance of the objects in the scene, and the spectral radiance of the illumination.*

Supposing a linear optoelectronic transfer function of the acquisition system, the camera response c_k for an image pixel is then equal to

$$
\begin{aligned}
c_k &= \int_{\lambda_{\min}}^{\lambda_{\max}} l_R(\lambda) r(\lambda) o(\lambda) \phi_k(\lambda) a(\lambda) \, d\lambda + \epsilon_k \\
&= \int_{\lambda_{\min}}^{\lambda_{\max}} r(\lambda) \omega_k(\lambda) \, d\lambda + \epsilon_k
\end{aligned}
\tag{6.1}
$$

where $\omega_k(\lambda) = l_R(\lambda) o(\lambda) \phi_k(\lambda) a(\lambda)$ denotes the spectral sensitivity of the kth channel, and ϵ_k is the additive noise. The assumption of system linearity comes from the fact that the CCD sensor is inherently a linear device. However, for real acquisition systems this assumption may not hold, for example due to electronic amplification non-linearities or stray light in the camera (Farrell and Wandell, 1993, Maître *et al.*, 1996). Then, appropriate nonlinear corrections may be necessary (see Sections 3.2.2 and 7.3).

By modeling the nonlinearities of the camera as

$$\check{c}_k = \Gamma\left(\int_{\lambda_{\min}}^{\lambda_{\max}} r(\lambda)\omega_k(\lambda)\,d\lambda + \epsilon_k\right), \tag{6.2}$$

cf. Eq. 6.1, we may easily obtain the response

$$c_k = \Gamma^{-1}(\check{c}_k) \tag{6.3}$$

of an ideal linear camera by inverting the function Γ.

By uniformly sampling[1] the spectra at N equal wavelength intervals, we can rewrite Eq. 6.1 as a scalar product in matrix notation as

$$c_k = \mathbf{r}^t \boldsymbol{\omega}_k + \epsilon_k, \tag{6.4}$$

where

$$\boldsymbol{\omega}_k = [\omega_k(\lambda_1)\,\omega_k(\lambda_2)\dots\omega_k(\lambda_N)]^t,$$

and

$$\mathbf{r} = [r(\lambda_1)\,r(\lambda_2)\dots r(\lambda_N)]^t,$$

are the vectors containing the spectral sensitivity of the kth channel of the acquisition system, and the sampled spectral reflectance, respectively.

6.2.2   Spectral sensitivity function estimation

Let us now consider a K channels acquisition system, the system unknowns of which are represented by the vectors $\boldsymbol{\omega}_k$, $k = 1 \dots K$. Two classes of methods exist for the estimation of these vectors, being referred to as the system's *spectral sensitivity functions*. The first class of methods is based on direct spectral measurements, requiring quite expensive equipment, in particular a wavelength-tunable source of monochromatic light. The camera characteristics is determined by individually evaluating the camera responses to monochromatic light from each sample wavelength of the visible spectrum (Park *et al.*, 1995, Burns, 1997, Martínez-Verdú *et al.*, 1998, Sugiura *et al.*, 1998).

The second type of approach is based on the acquisition of a number of samples with known reflectance or transmittance spectra. By observing the camera output to known input, we may estimate the camera sensitivity. Several authors have reported the use of such methods, *e.g.* Pratt and

---

[1] For a discussion of the sampling of color spectra, see Trussell and Kulkarni (1996).

Mancill (1976), Sharma and Trussell (1993; 1996c), Farrell and Wandell (1993), Sherman and Farrell (1994), Hubel *et al.* (1994), Burger and Sherman (1994), Farrell *et al.* (1994), Maître *et al.* (1996), Hardeberg *et al.* (1998b). We adopt this second approach.

To perform an estimation of ω_k for a given channel k, the camera responses $c_{k,p}, p = 1 \ldots P$, corresponding to a selection of P color patches with known reflectances $\mathbf{r}_p$ are measured. Denoting the sampled spectral reflectances of all the patches as the matrix $\mathbf{R} = [\mathbf{r}_1 \mathbf{r}_2 \ldots \mathbf{r}_P]$, the camera response of the kth channel to the pth sample as $c_{k,p}$, the response of the kth channel to these P samples, $\mathbf{c}_{k,P} = [c_{k,1} c_{k,2} \ldots c_{k,P}]^t$, is given by

$$\mathbf{c}_{k,P} = \mathbf{R}^t \omega_k + \epsilon_k. \tag{6.5}$$

For P target patches and a spectral sampling rate of N equally spaced wavelengths, the camera responses $\mathbf{c}_{k,P}$ and the noise term ϵ_k are vectors of P rows, $\mathbf{R}^t$ is a (P-rows $\times N$-columns) matrix, and ω_k is a vector of N rows.

When the reflectance spectra of the P target patches and the corresponding camera responses $\mathbf{c}_{k,P}$ are known, Equation 6.5 can be used as a basis for the estimation $\hat{\omega}_k$ of the spectral sensitivities ω_k. We present several methods for this estimation in the following sections.

6.2.2.1 Simulation setup

To illustrate the performance of the different estimation techniques, we have performed a simulation of an image acquisition system. The spectral sensitivity curves of an Eikonix color CCD camera as provided by the manufacturer were used, together with the CIE illuminant A (see Section 2.4.6) as shown in Figure 6.2 (the spectral transmittance of the optical path is assumed perfect: $o(\lambda) = 1$). The spectral reflectances of 1269 matte Munsell color chips, *cf.* Section 6.4.3, were used as input to the camera model. Following the scheme shown in Figure 6.3 we evaluate the quality of the spectral characterization methods by comparing the spectral sensitivity functions estimated by the methods, to the real functions as defined by our simulation.

6.2.2.2 Noise considerations

To evaluate the robustness of the different methods to noise, we have simulated noisy image acquisition process. But what kind of noise corresponds

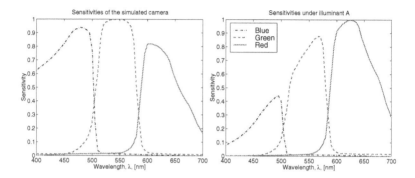

Figure 6.2: *Spectral sensitivities of the three channels of the simulated camera (left) and the resulting spectral sensitivities of the acquisition system using the illuminant A (right).*

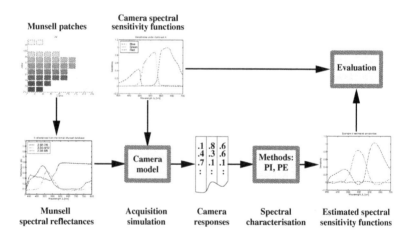

Figure 6.3: *Simulation of the acquisition system and evaluation of the spectral characterization methods, the Principal Eigenvector (PE) and Pseudoinverse (PI) methods, as described in the following sections.*

to real-life situations? A non-exhaustive list of possible error sources that may have influence on the spectral sensitivity estimation quality follows:

▓ Quantization noise

▓ Random noise in the acquisition system

▓ Reflectance spectra measurement errors

▓ Difference in viewing/illumination geometry between the image acquisition setup and the reflectance measurements using a spectrophotometer

▓ Deviation from linear acquisition model due to effects such as *i)* insufficiently corrected non-linear transfer function $\Gamma^{-1}(\cdot)$, *ii)* too coarse spectral sampling, *iii)* residual camera sensitivity outside of the wavelength interval used in the model, *iv)* fluorescence

In the literature several authors were concerned about acquisition noise. Hubel *et al.* (1994) showed that in a simulation with 1% acquisition noise, the PE method yielded rather poor results. Finlayson *et al.* (1998) also uses simulations with 1% noise. Sharma and Trussell (1996c) estimated the overall noise level in a commercial 8 bit/channel flatbed color scanner to be 33 dB. For an in-depth analysis of noise in multispectral image acquisition systems, refer to Burns (1997).

In our simulations, we have chosen to consider only the quantization noise, and we have simulated acquisitions using different numbers of bits for the data encoding. The camera responses are normalized before quantization so that the response of a perfect reflecting diffuser yields values of $c_k = 1.0, k = 1 \dots K$.

The relationship between the number of bits b used to encode the camera response, and the signal-to-noise ratio (SNR) is given by

$$\text{SNR [dB]} = 10 \log_{10} \left(\frac{\|\mathbf{c}_{k,P}\|^2}{\|\mathbf{c}_{k,P} - \text{quant}_b(\mathbf{c}_{k,P})\|^2} \right), \quad (6.6)$$

where $\text{quant}_b(\mathbf{c}_{k,P})$ represents the quantization/dequantization of $\mathbf{c}_{k,P}$ on b bits. Applied to the entire Munsell database, we obtain the SNRs given in Table 6.1.

Table 6.1: *Signal-to-noise ratio (SNR) for different number of bits used for quantization. (Mean values of the three channels SNR's, estimated on the entire Munsell spectral database, cf. Section 6.4.3.)*

Number of bits, b	4	5	6	7	8
SNR [dB]	25.0	31.0	37.4	43.3	49.6
	9	10	12	14	16
	55.5	61.6	73.7	85.6	97.5

6.2.2.3 Pseudoinverse (PI) estimation

Based on Equation 6.5, $\mathbf{c}_{k,P} = \mathbf{R}^t \boldsymbol{\omega}_k + \boldsymbol{\epsilon}_k$, we may estimate the system unknowns $\boldsymbol{\omega}_k$, $k = 1 \dots K$ from the following equation:

$$\hat{\boldsymbol{\omega}}_k = (\mathbf{R}\mathbf{R}^t)^{-1}\mathbf{R}\mathbf{c}_{k,P} = (\mathbf{R}^t)^{-}\mathbf{c}_{k,P}, \tag{6.7}$$

where $(\mathbf{R}^t)^{-}$ denotes the Moore-Penrose pseudoinverse (Albert, 1972) of $\mathbf{R}^t$ which, *in the absence of noise*, minimizes the root-mean-square (RMS) estimation error

$$d_E = \|\boldsymbol{\omega}_k - \hat{\boldsymbol{\omega}}_k\|. \tag{6.8}$$

If $\mathrm{rank}(\mathbf{R}) \geq N$ (requiring at least that $P \geq N$) and without noise, this solution is exact up to the working precision. Under real world conditions, however, this system inversion is not straightforward.

We have evaluated the influence of quantization noise on the pseudoinverse (PI) estimation as a function of the number of bits used for data encoding. We find that the error increases drastically as the number of bits decreases, and that, even when using 8 bits, the quality of the estimation is still poor, see Figs. 6.4 and 6.5. This result illustrates the fact that the unmodified pseudoinverse method is not suitable in the presence of noise.

6.2.2.4 Principal Eigenvector (PE) method

With the Principal Eigenvector (PE) method, also known as the rank-deficient pseudoinverse, several authors (Sharma and Trussell, 1993; 1996c, Hubel *et al.*, 1994, Farrell and Wandell, 1993, Hardeberg *et al.*, 1998b) reduced the noise sensitivity of the system inversion by only taking into account singular vectors corresponding to the most significant singular values. A Singular Value Decomposition (SVD) (Golub and Reinsch, 1971,

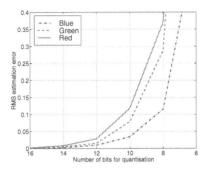

Figure 6.4: *RMS sensitivity estimation error for different numbers of bits using the pseudoinverse (PI) method.*

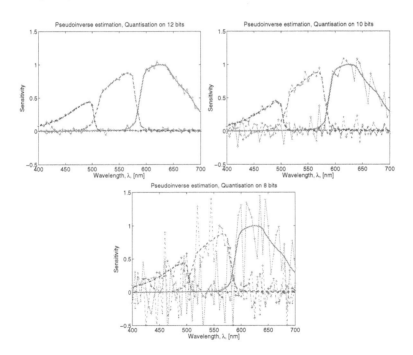

Figure 6.5: *Sensitivity estimation for different numbers of bits using the pseudoinverse (PI) method. The estimated sensitivities are drawn with thin lines. We see that the pseudoinverse method is very sensitive to noise.*

Jolliffe, 1986) is applied to the matrix $\mathbf{R}$ of the spectral reflectances of the observed patches.

We recall[2] that for any $(P \times N)$ matrix $\mathbf{X}$ of rank R, there exist a $(P \times P)$ unitary matrix $\mathbf{U}$ and an $(N \times N)$ unitary matrix $\mathbf{V}$ for which

$$\mathbf{X} = \mathbf{U}\mathbf{W}\mathbf{V}^t, \tag{6.9}$$

where $\mathbf{W}$ is a $(P \times N)$ matrix with general diagonal entry $w_i, i = 1 \ldots R$, called a *singular value* of $\mathbf{X}$, all the other entries of $\mathbf{W}$ being zero. The columns of the unitary matrix $\mathbf{U}$ are composed of the eigenvectors $\mathbf{u}_i$, $i = 1 \ldots P$, of the symmetric matrix $\mathbf{X}\mathbf{X}^t$. Similarly the columns of the unitary matrix $\mathbf{V}$ are composed of the eigenvectors $\mathbf{v}_j$, $j = 1 \ldots N$ of the symmetric matrix $\mathbf{X}^t\mathbf{X}$. Since $\mathbf{U}$ and $\mathbf{V}$ are unitary matrices, it can easily be verified that when $\mathbf{X} = \mathbf{U}\mathbf{W}\mathbf{V}^t$, *cf.* Equation 6.9, then $\mathbf{X}^- = \mathbf{V}\mathbf{W}^-\mathbf{U}^t$, where $\mathbf{W}^-$ has a general diagonal entry equal to w_i^{-1}, $i = 1 \ldots R$, and zeros elsewhere.

If the noise ϵ_k is assumed to be uncorrelated and signal independent, with a variance of $\sigma_{\epsilon_k}^2$, the expected estimation error depends upon the term $\sum_{i=1}^{R} \sigma_{\epsilon_k}^2 / w_i^2$ (Sharma and Trussell, 1993). It is clear that if any of the singular values are small, the error will be large. It has been found by several studies, *e.g.* Maloney (1986), Wandell (1987), Parkkinen *et al.* (1989), Vrhel *et al.* (1994), that the singular values of a matrix of spectral reflectances such as $\mathbf{R}^t$ are strongly decreasing, and by consequence that reflectance spectra can be described accurately by a quite small number of parameters. It has thus been proposed to only take into account the first $r < R$ singular values in the system inversion. The spectral sensitivity of the kth channel may thus be estimated by

$$\hat{\omega}_k = \mathbf{V}\mathbf{W}^{(r)-}\mathbf{U}^t\mathbf{c}_{k,P}, \tag{6.10}$$

where $\mathbf{W}^{(r)-}$ has a general diagonal entry equal to w_i^{-1}, $i = 1 \ldots r$, $(r < R)$, and zeros elsewhere.

For the PE method we have evaluated the influence of the choice of the number of eigenvectors, r, that is to be considered as principal eigenvectors, for different levels of quantization noise. From Figure 6.6 we conclude that the estimation error is due to two factors, *a)* the inability of the model of reduced dimension to fit the sensitivity curves, and *b)* the

---

[2] See Appendix A.3 for a more thorough presentation of the singular value decomposition algorithm.

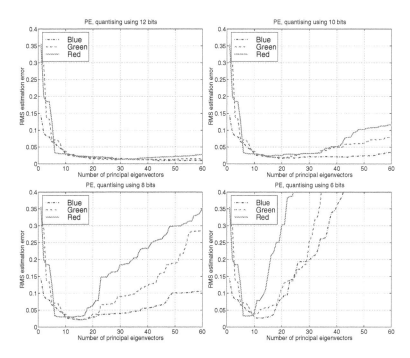

Figure 6.6: *RMS sensitivity estimation error for different levels of quantization noise. We see that too few and too many principal eigenvectors causes high estimation errors.*

noise, being dominant when a greater number of eigenvectors are used in the computations. When the noise is increased, this error becomes predominant for lower dimensions. The optimal numbers of PE's to use for different levels of acquisition noise are shown in Figure 6.7, together with the resulting RMS estimation errors using the optimal number of PE's. We conclude that if the amount of noise increases, fewer eigenvectors should be used for the spectral sensitivity estimation.

In Figure 6.8 we present some examples of sensitivity estimation using the PE method with $r = 10$ and $r = 20$ principal eigenvectors, and quantizing on 10 and 8 bits. We see that when using 10 principal eigenvectors, the quantization error has little influence on the estimation quality, whereas when 20 principal components are used, and when quantizing on 8 bits, the quantization noise severely deteriorates the estimation results.

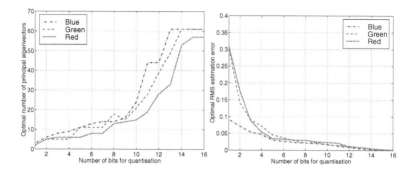

Figure 6.7: *Optimal number of PE's to use for the inversion for different levels of acquisition noise (left) and the resulting RMS estimation errors using the optimal number of PE's (right). The optimal numbers are defined by the minima in Figure 6.6. As the amount of noise increases, fewer principal eigenvectors should be used for the spectral sensitivity estimation. If the noise is too high, good results cannot be obtained.*

6.2.2.5   Selection of the most significant target patches

In order to yield reliable results, the principal eigenvector method as described in the previous section requires to use a very large number of target patches; this is a severe limitation for practical applications, in which one would seek to have, *e.g.*, 20 patches rather than 1000. To solve this problem, we propose the following method for the selection of those reflectance samples $\mathbf{r}_{s_1}, \mathbf{r}_{s_2}, \mathbf{r}_{s_3}, \ldots$ which are most significant for the estimation of the camera's spectral sensitivity.

Starting from the full set of all available spectral reflectance functions $\mathbf{r}_p, p = 1 \ldots P$, we first select that $\mathbf{r}_{s_1}$ which is of maximum RMS value:

$$\|\mathbf{r}_{s_1}\| \geq \|\mathbf{r}_p\| \quad \text{for } p = 1 \ldots P \qquad (6.11)$$

Next, we select that $\mathbf{r}_{s_2}$ which minimizes the *condition number* (Matlab Language Reference Manual, 1996) of $[\mathbf{r}_{s_1} \mathbf{r}_{s_2}]$, that is, the ratio of the largest to the smallest singular value. Denoting $w_{\min}(\mathbf{X})$ and $w_{\max}(\mathbf{X})$ the minimum and maximum singular values of a matrix $\mathbf{X}$, this minimization

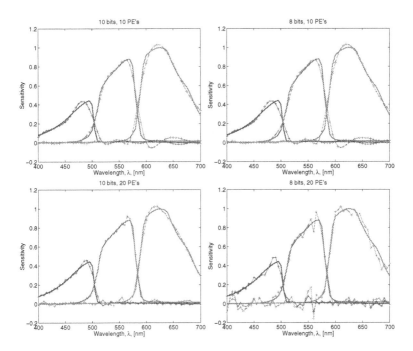

Figure 6.8: *Camera sensitivity estimation using the PE method with 10 (upper) and 20 (lower) principal eigenvectors, and quantizing on 10 (right) and 8 (left) bits. We see that when quantizing on 8 bits the quantization noise deteriorates the estimation results severely when using too many principal eigenvectors in the estimation, cf. Fig. 6.6.*

may be expressed by the following expression:

$$\frac{w_{\max}([\mathbf{r}_{s_1}\mathbf{r}_{s_2}])}{w_{\min}([\mathbf{r}_{s_1}\mathbf{r}_{s_2}])} \leq \frac{w_{\max}([\mathbf{r}_{s_1}\mathbf{r}_p])}{w_{\min}([\mathbf{r}_{s_1}\mathbf{r}_p])},$$

$$\text{for } p = 1\ldots P,\ p \neq s_1. \tag{6.12}$$

Further sample spectra are added according to the same rule, *i.e.* for the choice of the ith sample $\mathbf{r}_{s_i}$:

$$\frac{w_{\max}([\mathbf{r}_{s_1}\mathbf{r}_{s_2}\ldots\mathbf{r}_{s_i}])}{w_{\min}([\mathbf{r}_{s_1}\mathbf{r}_{s_2}\ldots\mathbf{r}_{s_i}])} \leq \frac{w_{\max}([\mathbf{r}_{s_1}\mathbf{r}_{s_2}\ldots\mathbf{r}_{s_{i-1}}\mathbf{r}_p])}{w_{\min}([\mathbf{r}_{s_1}\mathbf{r}_{s_2}\ldots\mathbf{r}_{s_{i-1}}\mathbf{r}_p])},$$

$$\text{for } p = 1\ldots P,\ p \notin \{s_1, s_2, \ldots s_{i-1}\}. \tag{6.13}$$

By this procedure, we obtain a set of most significant reflectance samples, for the spectral sensitivity estimation process. The motivation behind this method is to choose, for each iteration step, a reflectance spectrum that is as "different" as possible from the other target spectra.

In order to evaluate the target patch selection method we compared the results obtained from "step by step" optimally chosen target samples with those obtained from a heuristically chosen set of samples. The heuristically chosen set was obtained by simply selecting the patch of highest chroma from each of the 20 hue angle pages of the Munsell atlas. The Munsell notations[3] and CIELAB a^* and b^* coordinates for both the optimally chosen set and the heuristically chosen set are shown in Figure 6.9.

We estimated the camera spectral sensitivities independently from both sets, for 8-bit, 10-bit and 12-bit quantization, using the PE method with 10 and 20 principal eigenvectors, denoted PE(10) and PE(20), respectively. Figure 6.10 presents the estimated spectral sensitivities with 10-bit quantization, on the top for the heuristically chosen set, and in the middle for the optimally chosen set. The comparison between the upper and middle panels of Figure 6.10 shows that the results obtained from the optimal set of target patches present a much better fit to the camera's spectral sensitivities than the results obtained from the heuristically chosen set. In the lower part of the figure we present the estimations using the *Macbeth ColorChecker* (McCamy *et al.*, 1976), a color target which is extensively used for color calibration tasks (see Section 7.2.4 and *e.g.* Farrell and Wandell, 1993, Finlayson *et al.*, 1998, Burns, 1997).

---

[3]The Munsell color specification is given in the order Hue, Value, Chroma, for example, 5R 4/14 is on the principal red axis, with a lightness slightly darker than a medium gray, and that it has a very strong chroma. See *e.g.* Hunt (1991), Section 7.4, for more details.

Optimal	Heuristic
7.5 RP 9/2	5 R 5/14
5 R 4/14	10 R 6/12
7.5 Y 8/12	5Y R 7/12
2.5 G 7/10	10 YR 7/12
5 P 2.5/6	5 Y 8/12
10 R 7/12	10 Y 8/12
7.5 RP 6/10	5 GY 8.5/10
2.5 B 5/8	10 GY 7/10
10 P 3/8	5 G 7/10
7.5 R 7/4	10 G 6/10
10 B 6/10	5 BG 6/8
10 Y 8/4	10 BG 6/8
7.5 YR 8/8	5 B 6/8
10 RP 8/6	10 B 6/10
10 R 3/2	5 PB 5/12
7.5 PB 5/12	10 PB 5/10
10 Y 8.5/6	5 P 5/10
10 PB 4/10	10 P 5/12
10 YR 3/1	5 RP 5/12
7.5 YR 6/4	10 RP 5/12

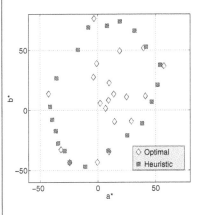

Figure 6.9: *The Munsell notations (left) and the CIELAB a^* and b^* coordinates (right) for the Munsell patches chosen either heuristically or optimally by the proposed method.*

We see from Figure 6.10 that the results are clearly better for the optimally chosen reflectances than when using the heuristically selected and Macbeth patches. These results are in good accordance with the results obtained from a Principal Component Analysis (PCA) of the three reflectance sets using the methods that we will present in Section 6.4. In Figure 6.11 we compare the magnitudes of the singular values, and wee see that the effective dimension D_e of the optimally selected Munsell patches is generally higher than for the two other sets, while the condition number is smaller, see Table 6.2. We can also conclude from this analysis that great attention must have been given to spectral properties when the Macbeth chart was designed (McCamy *et al.*, 1976).

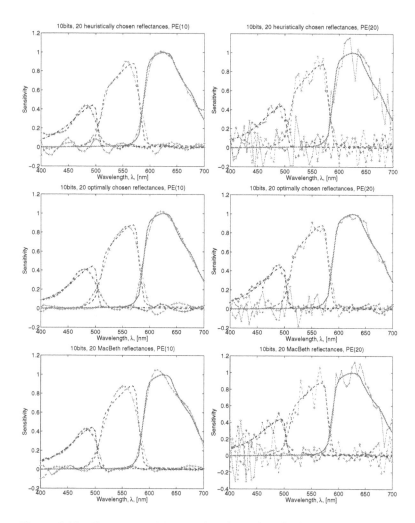

Figure 6.10: *Camera sensitivity estimation using the PE method with 10 (left) and 20 (right) principal eigenvectors, with 20 heuristically chosen reflectances (upper), 20 optimally chosen reflectances (middle), and 20 Macbeth reflectances (lower). We see that we attain much better results when choosing reflectance samples using a vectorspace approach than when choosing in a heuristic manner or using the Macbeth ColorChecker.*

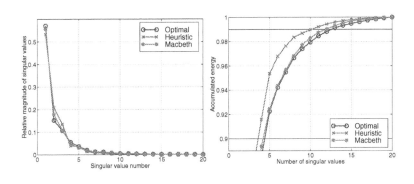

(a) Relative magnitude of singular values

(b) Accumulated energy E_a

Figure 6.11: *PCA analysis of the optimally selected subset of 20 Munsell patches, of the heuristically selected ones, and of 20 patches of the Macbeth ColorChecker. (Four medium gray patches out of the 24 Macbeth patches are excluded from the data set to enable comparison with the same number of color patches.)*

Table 6.2: *Comparison of the results of a PCA analysis of the optimal/heuristic selections of Munsell patches and of the Macbeth ColorChecker.*

	OPTIMAL	HEURISTIC	MACBETH
Effective dimension D_e for $E_{req} = 0.90$	5	4	5
Effective dimension D_e for $E_{req} = 0.99$	13	11	12
Condition number	672	2274	2053

## 6.2.3	Discussion on spectral characterization

In order to compare the results of the spectral sensitivity estimation using the different methods proposed in the previous sections, we present in Table 6.3 the RMS estimation errors for quantization using 8, 10 and 12 bits, using either all 1269 Munsell reflectances, 20 heuristically chosen reflectances, or 20 optimally chosen reflectances. In Table 6.4 we present the estimation errors obtained with the best method for each noise level, and for each of the selected data sets, relatively to the results obtained with the complete set of Munsell chips. Several conclusions may be drawn from these results.

▧ A coarser quantization always increases the spectral sensitivity estimation error.

▧ The unmodified pseudoinverse (PI) method is generally not suitable in the presence of noise. The Principal Eigenvector (PE) method should be used instead.

▧ The optimal number of principal eigenvectors used in the PE method depends on the noise level.

▧ The results are always better when using "step by step" optimally chosen samples than when using samples chosen heuristically.

▧ The quality of the estimation is almost as good using 20 optimally chosen samples as when using the complete set of 1269 Munsell chips. In a practical situation, this is of great importance, as it decreases drastically the workload needed to perform a spectral characterization of electronic cameras.

We find that the obtained spectral sensitivity estimations are of rather high accuracy, for moderate levels of noise. However, the results could be further improved by adding constraints such as smoothness and positivity to the estimation methods. For example the Wiener estimation as used by Pratt and Mancill (1976) and Hubel et al. (1994) or the technique of projection onto convex sets (POCS), proposed by Sharma and Trussell (1993; 1996c) could be used.

When considering illuminants having spiky spectral radiances, such as fluorescent lamps, the spectral sensitivity estimation becomes more difficult than in our case, where the illuminant A is used. The use of fluorescent

Table 6.3: *Root-mean-square spectral sensitivity estimation errors using the different methods. For simplicity, the mean values for the three channels are used. We see that the quality of the estimation is almost as good using 20 optimally chosen samples than when using the complete set of 1269 Munsell chips. See also Table 6.4.*

		8 bit	10 bit	12 bit
All 1269 reflectances	PI	0.25797	0.07752	0.01800
	PE(20)	0.04350	0.02027	0.01796
	PE(10)	0.03178	0.03171	0.03170
20 optimally chosen reflectances	PE(20)=PI	0.19568	0.05365	0.02498
	PE(10)	0.04712	0.03821	0.03772
20 heuristically chosen reflectances	PE(20)=PI	0.40801	0.10726	0.04472
	PE(10)	0.04734	0.04261	0.04159

Table 6.4: *Estimation errors for the optimal/heuristic data sets relative to the results obtained when using the complete set of 1269 Munsell chips. For each of the data sets, the best method of Table 6.3 is used. We see that the difference between the optimally and heuristically chosen sets becomes larger for lower levels of noise.*

	8 bit	10 bit	12 bit
All	100	100	100
Optimal	148	189	139
Heuristic	149	210	232

lamps in desktop scanners thus presents a severe difficulty for the spectral characterization. It could be considered to represent the spectra by a combination of smooth basis vectors and ray spectra. Furthermore, it should be emphasized that for real-life acquisitions, quantization noise is only one of many sources of error, as discussed in Section 6.2.2.2. Thus the quality of the estimation is likely to be poorer than in the simulations, for the same number of bits.

6.3 Spectral reflectance estimation from camera responses

We will now consider a multispectral image capture system consisting of a monochrome CCD camera and a set of K color filters, for a given illuminant. The spectral characteristics $\omega(\lambda)$ of the image acquisition system including the illuminant but without the filters is supposed known, cf. Section 6.2. The known spectral transmittances of the filters are denoted by $\phi_k(\lambda)$, $k = 1 \ldots K$. Analogously to Equation 6.1 the camera response c_k obtained with the kth filter, discarding acquisition noise, is given by

$$c_k = \int_{\lambda_{\min}}^{\lambda_{\max}} r(\lambda)\phi_k(\lambda)\omega(\lambda)d\lambda. \qquad (6.14)$$

The vector $\mathbf{c}_K = [c_1 c_2 \ldots c_K]^t$ representing the response to all K filters may be described using matrix notation as

$$\mathbf{c}_K = \Theta^t \mathbf{r}, \qquad (6.15)$$

where Θ is the known N-line, K-column matrix of filter transmittances multiplied by the camera characteristics, that is $\Theta = [\phi_k(\lambda_n)\omega(\lambda_n)]$.

We now address the problem of how to retrieve spectrophotometric information from these camera responses. A first approach is to define a direct colorimetric transformation from the camera responses $\mathbf{c}_K$ into for example the CIELAB space (see Section 2.4.8.1), under a given illuminant, minimizing typically the RMS error in a way similar to what is often done for conventional three-channel image acquisition devices, cf. Section 3. Given an appropriate regression model, this is found to give quite satisfactory results in terms of colorimetric errors (Burns, 1997). However, for our applications we are often concerned not only with the colorimetry of the imaged scene, but also with the inherent surface spectral reflectance of the viewed objects. Thus the colorimetric approach is not always sufficient.

In existing multispectral acquisition systems, the filters often have similar and rather narrow bandpass shape and are located at approximately equal wavelength intervals. For the reconstruction of the spectral reflectance, it has been proposed to apply interpolation methods such as spline interpolation or Modified Discrete Sine Transformation (MDST) (Keusen and Praefcke, 1995, Keusen, 1996). Such methods are not well adapted to filters having more complex wide-band responses, and suffer from quite severe aliasing errors (Burns, 1997, König and Praefcke, 1998a).

We adopt a linear-model approach and formulate the problem of the estimation of a spectral reflectance $\tilde{r}$ from the camera responses c_K as finding a matrix Q that reconstructs the spectrum from the K measurements as follows

$$\tilde{r} = Q c_K. \tag{6.16}$$

Our goal will thus be to determine a matrix Q that minimizes a distance $d(r, \tilde{r})$, given an appropriate error metric d. Some solutions to this problem are presented and discussed in the following sections.

6.3.1   Pseudo-inverse solution

An immediate solution[4] for estimating the spectral reflectance consists in simply inverting Equation (6.15) by using a pseudo-inverse approach, which provides us with the following minimum norm solution (Albert, 1972)

$$\tilde{r} = (\Theta \Theta^t)^{-1} \Theta c_K = (\Theta^t)^- c_K. \tag{6.17}$$

The pseudo-inverse reconstruction operator Q_0 is thus given by

$$Q_0 = (\Theta)^{t-}. \tag{6.18}$$

If the matrix Θ were of full rank N, and if we assume noiseless recordings, this method of reconstruction would be perfect. However it is not very well adapted in practical situations.[5] First, in order to achieve that the rank of Θ equals N, the number of color filters K should be at least equal to the number of spectral sampling points N. Furthermore this representation is very sensitive to signal noise. In fact, by this solution we minimize

---

[4] As implemented by Camus-Abonneau and Camus (1989)
[5] This was observed by Goulam-Ally (1990b) and Nagel (1993)

the Euclidian distance $d_E(\Theta^t \mathbf{r}, \mathbf{c}_K)$ in the camera response domain. A small distance does not guarantee the spectra $\mathbf{r}$ and $\tilde{\mathbf{r}}$ to be close, only that their projections into camera response space are close. Nevertheless, this approach is used by Tominaga (1996; 1997) to recover the spectral distribution of the illuminant from a six-channel acquisition. However, he applies a nested regression analysis to choose the proper number of components in order to better describe the spectrum and to increase the spectral-fit quality.

6.3.2 Reconstruction exploiting *a priori* knowledge of the imaged objects

We now define another reconstruction operator $\mathbf{Q}_1$ that minimizes the Euclidian distance $d_E(\mathbf{r}, \tilde{\mathbf{r}})$ between the original spectrum $\mathbf{r}$ and the reconstructed spectrum $\tilde{\mathbf{r}} = \mathbf{Q}_1 \mathbf{c}_K$. To achieve this minimization we take advantage of *a priori* knowledge on the spectral reflectances that are to be imaged. We know that the spectral reflectances of typical objects are smooth. We present this by assuming that the reflectance in each pixel is a linear combination of a set of smooth *basis functions*. We will typically use a set of *measured* spectral reflectances as basis functions, but other sets of functions could be used, *e.g.* a Fourier basis. Denoting the basis function "reflectances" as $\mathbf{R} = [\mathbf{r}_1 \mathbf{r}_2 \ldots \mathbf{r}_P]$, our assumption implies that, for any observed reflectance $\mathbf{r}$, a vector of coefficients $\mathbf{a}$ exists[6] such that any reflectance $\mathbf{r}$ may be expressed as

$$\mathbf{r} = \mathbf{R}\mathbf{a}. \qquad (6.19)$$

Hence, we obtain $\tilde{\mathbf{r}}$ from $\mathbf{a}$ by using Equations 6.16, 6.15, and 6.19:

$$\tilde{\mathbf{r}} = \mathbf{Q}_1 \mathbf{c}_K = \mathbf{Q}_1 \Theta^t \mathbf{r} = \mathbf{Q}_1 \Theta^t \mathbf{R} \mathbf{a}. \qquad (6.20)$$

With Equations 6.19 and 6.20 the ideal expression $\mathbf{r} = \tilde{\mathbf{r}}$ becomes

$$\mathbf{Q}_1 \Theta^t \mathbf{R} \mathbf{a} = \mathbf{R} \mathbf{a}. \qquad (6.21)$$

Assuming that $\mathbf{R}$ is a statistically significant representation of the reflectances that will be encountered for a given application, Equation 6.21 should be true for any $\mathbf{a}$, and hence

$$\mathbf{Q}_1 \Theta^t \mathbf{R} = \mathbf{R}. \qquad (6.22)$$

---

[6]See Bournay (1991).

This gives then the reconstruction operator minimizing the RMS spectral error by a pseudo-inverse approach as

$$\mathbf{Q}_1 = \mathbf{RR}^t\Theta(\Theta^t\mathbf{RR}^t\Theta)^{-1}. \qquad (6.23)$$

The choice of the spectral reflectances in $\mathbf{R}$ should be well representative of the spectral reflectances encountered in the applications. In our experiments on paintings we used a set of 64 spectral reflectances of pure pigments used in oil painting and provided to us by the National Gallery in London (Maître *et al.*, 1996). For other applications, we could use sets that are supposed to be representative of general reflectances, such as the object colors of Vrhel *et al.* (1994) or the natural colors of Jaaskelainen *et al.* (1990), see Section 6.4.

Note that slightly different methods exist for the estimation of a spectral reflectance from the camera responses, such as the Wiener estimation method (König and Praefcke, 1998a, Vrhel and Trussell, 1994, Haneishi *et al.*, 1997) based on the autocorrelation matrix of $\mathbf{R}$, and a principal component analysis method where the principal components (*cf.* Appendix A.2) of the spectral reflectance are estimated by a least mean square approach from the camera responses (Burns, 1997).

6.3.3 Evaluation of the spectral reflectance reconstruction

We have performed a rapid evaluation of the two reconstruction operators presented in Sections 6.3.1 and 6.3.2. The experimental results shown in Figure 6.12 indicate clearly that the reconstruction operator $\mathbf{Q}_1$ is much better than $\mathbf{Q}_0$, as expected.

6.4 Analysis of spectral reflectance data sets

As noted in Section 6.3 it may be of strong interest to have knowledge about the nature of the reflectance spectra which are of interest for an image acquisition. The effective dimension of reflectance spectra, that is, the number of components needed to describe a spectrum in a vectorial space is discussed extensively in the literature, see Appendix F.

For Munsell colors, Cohen (1964) states that three components is sufficient, Eem *et al.* (1994) propose four, Maloney (1986) proposes five to

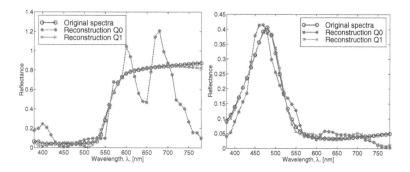

Figure 6.12: *Reconstruction of the spectral reflectance of the cadmium orange (left) and the manganese blue (right) pigments from the camera responses using seven filters. Note the clear superiority of using operator* Q_1*, which takes into account a priori knowledge on the spectral reflectances encountered in oil painting.*

seven, Burns propose five (Burns and Berns, 1996) or six (Burns, 1997), Lenz *et al.* (1995; 1996a) use six and Parkkinen *et al.* (1989) and Wang *et al.* (1997) say eight.

For data including natural reflectances, Dannemiller (1992) states that three is sufficient, Vrhel *et al.* (1994) propose three to seven basis functions, Praefcke (1996) propose five, while Keusen (1996) states that up to ten is needed.

For human skin, three components is proposed by Imai *et al.* (1996a;b). For oil painting, five is proposed by Tsumura and coworkers (Haneishi *et al.*, 1997, Yokoyama *et al.*, 1997), while Maître *et al.* (1996) claims that ten to twelve factors are needed. García-Beltrán *et al.* (1998) use a linear basis of seven vectors to represent the spectral reflectance of acrylic paints.

For a practical realization of a multispectral scanner using filters, these results should be taken into account when designing the system, in particular when choosing the number of acquisition channels. Also here, the choices found in the literature are many: three (Imai *et al.*, 1996a;b), (Shiobara *et al.*, 1995; 1996), four (Chen and Trussell, 1995), five (Haneishi *et al.*, 1997, Yokoyama *et al.*, 1997, Kollarits and Gibbon, 1992), six (Tominaga, 1996; 1997), seven (Saunders and Cupitt, 1993, Martinez *et al.*, 1993, Burns and Berns, 1996), five to ten (Hardeberg *et al.*, 1998a; 1999), ten to twelve (Maître *et al.*, 1996), twelve to fourteen (Keusen and Praefcke, 1995, Keusen, 1996), and sixteen (König and Praefcke, 1998b).

We see that the existing conclusions concerning the dimension of spectral reflectances, and the number of channels needed to acquire multispectral images, are rather dispersed. Often the applied statistical analysis is quite elementary, and conclusions are drawn without clearly defined objectives.

We will thus proceed to a comparative statistical analysis of sets of spectral reflectances, using tools such as the Principal Component Analysis. We define some design criterions, and apply the analysis to different sets of reflectances.

6.4.1 Principal Component Analysis

A Principal Component Analysis (PCA), see Section A.2, is applied to the data set of reflectances. The reason for applying such an analysis is mainly twofold.

■ To acquire information about the dimensionality of the data. Are all the N reflectances linearly independent? This may give an indication on the number of filters to be chosen for the acquisition, see Section 6.5.

■ To allow a compression of the spectral information. A reflectance spectrum can be represented approximately by a reduced number of principal components.

To implement this analysis, we apply the Singular Value Decomposition (SVD), see Section A.3. Numerous variants of the SVD algorithm exist. We apply here the version presented in Section A.3.2, as implemented in Matlab (Matlab Language Reference Manual, 1996). We recall that for any arbitrary $(N \times P)$ matrix $\mathbf{X}$ of rank R, there exist an $(N \times N)$ unitary matrix $\mathbf{U}$ and a $(P \times P)$ unitary matrix $\mathbf{V}$ for which

$$\mathbf{X} = \mathbf{U}\mathbf{W}\mathbf{V}^t, \tag{6.24}$$

where $\mathbf{W}$ is an $(N \times P)$ matrix with general diagonal entries w_i, $i = 1 \ldots R$, denoted *singular values*[7] of $\mathbf{X}$, and the columns of the unitary matrix $\mathbf{U}$ are composed of the eigenvectors $\mathbf{u}_i$, $i = 1 \ldots N$, of the symmetric matrix $\mathbf{XX}^t$, *cf.* Section A.3.

---

[7]The *singular values* of $\mathbf{X}$ correspond to the square roots of the *eigenvalues* of $\mathbf{XX}^t$.

For our PCA analysis, we denote our data set of P reflectances $\mathbf{r}_i$ as the N-line, P-column matrix $\mathbf{R} = [\mathbf{r}_1 \mathbf{r}_2 \dots \mathbf{r}_P]$, and we define the matrix $\mathbf{X}$ as

$$\mathbf{X} = \begin{bmatrix} \mathbf{r}_1 - \bar{\mathbf{r}} & \mathbf{r}_2 - \bar{\mathbf{r}} & \cdots & \mathbf{r}_P - \bar{\mathbf{r}} \end{bmatrix}, \tag{6.25}$$

where $\bar{\mathbf{r}}$ contains the mean values of the reflectance spectra,

$$\bar{\mathbf{r}} = \frac{1}{P} \sum_{j=1}^{P} \mathbf{r}_j. \tag{6.26}$$

Applying the SVD, the resulting matrix $\mathbf{U}$ (*cf.* Eq. 6.24) is a $(N \times N)$ column-orthogonal matrix describing an orthogonal basis of the space spanned by $\mathbf{X}$ or equivalently by $\mathbf{R}$. We denote this space by $R(\mathbf{R})$, the *range* of $\mathbf{R}$, and note that $R(\mathbf{R}) = R(\mathbf{X}) = R(\mathbf{U})$. We will denote the columns of $\mathbf{U}$, called *nodes* in PCA terminology, as the *characteristic reflectances* of the data set. $\mathbf{W}$ is a diagonal matrix containing on its diagonal the singular values w_i, in order of decreasing magnitude,

$$\mathbf{W} = \begin{bmatrix} w_1 & & 0 \\ & \ddots & \\ 0 & & w_N \end{bmatrix}. \tag{6.27}$$

An example of such singular values for the case of a set of oil pigments as presented in Section 6.4.3 is shown in Figure 6.13.

We note that there is a strong concentration of variance/energy in the first few singular values. This suggests that the spectral reflectances may be approximated quite correctly using a small number of components, as described in the following section.

6.4.2 Effective dimension

The dimension D of the space $R(\mathbf{R})$ is rigorously determined by $D = \text{rank}(\mathbf{R})$, which is given by the number of non null singular values. If the columns of $\mathbf{R}$, the reflectance spectra, are linearly independent, then $D = N$.[8] However, if some singular values are very close to zero, which is often the case, *cf.* Figure 6.13 on the facing page, the *effective* dimension D_e of the space may be much smaller. That is, it is possible to represent the

---

[8]This supposing that $N < P$. If, however, $N > P$, then $D = P$.

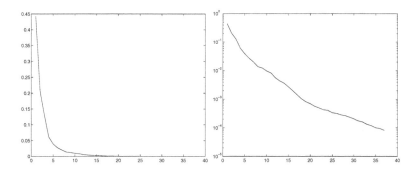

(a) Normalized singular values using a linear scale

(b) Normalized singular values using a logarithmic scale

Figure 6.13: *An example of the singular values of the matrix* **R** *containing the reflectance spectra of the pigments of the National Gallery chart. We note that there is a strong concentration of variance/energy in the first few singular values.*

spectral data in a more compact form, using merely D_e principal components, D_e being generally significantly smaller than D. Given a reflectance spectrum represented by the N-vector $\mathbf{x} = \mathbf{r} - \bar{\mathbf{r}}$, the vector of principal components $\mathbf{z} = [z_1 z_2 \ldots z_{\tilde{P}}]^t$ is given by (*cf.* Equation A.36)

$$\mathbf{z} = \tilde{\mathbf{U}}^t \mathbf{x}, \qquad (6.28)$$

$\tilde{\mathbf{U}}$ being defined as the first $\tilde{P} < P$ characteristic reflectances, $\mathbf{z}$ thus being a $\tilde{P}$-vector. The reconstruction of an approximation $\tilde{\mathbf{r}}$ of the original reflectance is obtained by (*cf.* Equation A.38)

$$\tilde{\mathbf{x}} = \tilde{\mathbf{U}}\mathbf{z} = \tilde{\mathbf{U}}\tilde{\mathbf{U}}^t \mathbf{x}, \qquad (6.29)$$

and consequently

$$\tilde{\mathbf{r}} = \tilde{\mathbf{U}}\mathbf{z} + \bar{\mathbf{r}} = \tilde{\mathbf{U}}\tilde{\mathbf{U}}^t(\mathbf{r} - \bar{\mathbf{r}}) + \bar{\mathbf{r}}. \qquad (6.30)$$

The spectral reconstruction error is thus identified as

$$d_E = \|\mathbf{r} - \tilde{\mathbf{r}}\| = \|\mathbf{r} - \tilde{\mathbf{U}}\tilde{\mathbf{U}}^t(\mathbf{r} - \bar{\mathbf{r}}) - \bar{\mathbf{r}}\| = \|\mathbf{x} - \tilde{\mathbf{U}}\tilde{\mathbf{U}}^t \mathbf{x}\| \qquad (6.31)$$

To determine an estimation of the effective dimension of the space $R(\mathbf{R})$, that is, a good choice of D_e, we need to determine how many principal components that must be taken into account to represent the data. To do this, in addition to the measurement of spectral reconstruction errors, the notion of *accumulated* energy $E_a(\tilde{P})$, that is the amount of the total energy, or signal variance, that is represented by the first $\tilde{P}$ singular vectors, turns out to be useful:

$$E_a(\tilde{P}) = \frac{\sum_{i=1}^{i=\tilde{P}} w_i}{\sum_{i=1}^{i=P} w_i}. \tag{6.32}$$

We may also define the *residual* energy $E_r(\tilde{P}) = 1 - E_a(\tilde{P})$, that is, the energy represented by the principal components that are not taken into account. As an example, we present in Figure 6.14 the mean and maximal spectral reconstruction error $\|\mathbf{r} - \tilde{\mathbf{r}}\|$, over the spectral reflectances of the base, compared to the residual energy, using $\tilde{P}$ principal components. We see that the mean spectral reconstruction error is highly correlated to the residual energy, while the maximal error shows a more random variation, although still correlated.

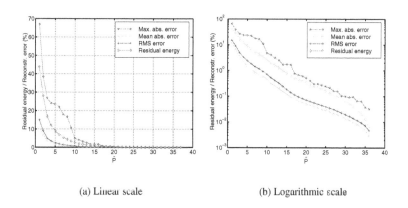

(a) Linear scale   (b) Logarithmic scale

Figure 6.14: *The mean and maximal spectral reconstruction error compared to the energy contained in the $\tilde{P}$ first principal components. Example is using the oil pigment data.*

This result suggests that we may use the accumulated (or equivalently the residual) energy as a criterion to define an appropriate choice of dimensionality D_e. We define thus

$$D_e \overset{\text{def}}{=} \min\{\tilde{P} | E_a(\tilde{P}) \geq E_{\text{req}}\} \tag{6.33}$$

The definition of the effective dimension depends thus on the choice of required accumulated energy E_{req}.

An example value seen in literature (Haneishi *et al.*, 1997, Burns, 1997) is $E_{\text{req}} = 99\%$, a choice which would give $D_e = 16$ and a RMS error of 0.14% for the data in the example shown in Figure 6.14. A choice of $D_e = 12$ as proposed by Maître *et al.* (1996) gives $E_a = 97.9\%$.[9]

6.4.3 Application to real reflectance data sets

We have chosen to apply our analysis to five distinct sets of reflectance spectra, examples of which are found in Figure 6.15.

1. PIGMENTS. We dispose of a significant set of color patches provided by the National Gallery in London (courtesy of David Saunders), providing the essential oil pigments used in the restoration of old paintings. This set contains $P = 64$ pure pigments covering all the shades of colors of the spectrum. The patches have been measured with a PhotoResearch PR-704 spectrophotometer on a wavelength interval of 380 nm - 780 nm, and a wavelength resolution of 2 nm.

2. MUNSELL. Thanks to Parkkinen *et al.* (1989) and their group at the University of Kuopio, Finland, we have access[10] to the reflectance spectra of 1269 matt Munsell color chips, (Munsell, 1976), measured using a Perkin-Elmer lambda 9 UV/VIS/NIR spectrophotometer, on a wavelength interval of 380 nm - 800 nm, and a wavelength resolution of 1 nm.

3. NATURAL. From the same source, we dispose of the reflectance spectra of 218 colored samples collected from the nature (Jaaskelainen *et al.*, 1990). Samples include flowers and leaves and other colorful plants. Each spectrum consists of 61 elements that are raw

---

[9]Note that a slightly different result ($E_a = 98.2\%$) was found in (Maître *et al.*, 1996) since the PCA/SVD analysis was performed on non-centered data.

[10]http://www.it.lut.fi/research/color/database/database.html

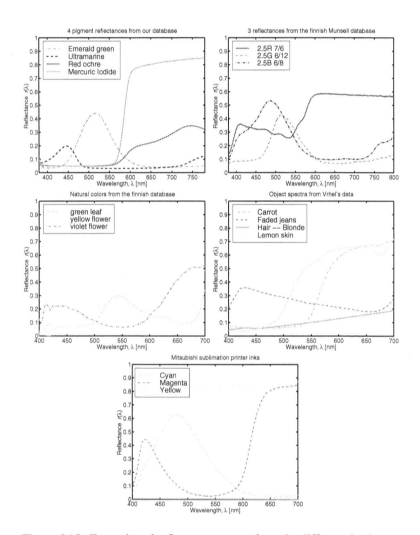

Figure 6.15: *Examples of reflectance spectra from the different databases:* PIGMENTS, MUNSELL, NATURAL, OBJECT, and SUBLIM.

data got from the output of the 12 bit A/D-converter of an Acousto-Optical Tunable Filters (AOTF) color measuring equipment, this corresponding to a wavelength interval of 400 nm - 700 nm, and a wavelength resolution of 5 nm.

4. OBJECT. Provided by Vrhel *et al.* (1994) and generously made available online[11], we dispose of 170 reflectance spectra from various natural and man-made objects measured by a PhotoResearch PR-703 spectroradiometer, and postprocessed to 10 nm intervals in the range from 400 nm - 700 nm.

5. SUBLIM. We have also measured using the PR-704 spectrophotometer a set of 125 patches of a Mitsubishi S340-10 sublimation printer. The patches are equally spaced in the printer CMY space.

To be able to compare these different reflectance spectrum databases, and to avoid the handling of extensively large matrices, we have resampled all data to a common wavelength resolution of 10nm, and a wavelength interval from 400nm - 700nm.

We then apply the PCA analysis as described in Section 6.4.1 to the different databases. The relative magnitude of the singular values, that is the eigenvalues of the covariance matrix $\mathbf{XX}^t$ are shown in Figure 6.16. In Figure 6.17 and Table 6.5, the accumulated energy represented by the $\tilde{P}$ first characteristic vectors is shown.

Analyzing the data reported in Table 6.5 we may conclude that a different number of basis vectors should be chosen, depending on the database used to calculate the covariance matrix, see Table 6.6. As expected, the NATURAL data shows the highest complexity, and the SUBLIM data the lowest, but they do exhibit a rather similar behavior. If, for example we require that $E_{req} = 99\%$ of the signal variance should be accounted for, we can encode the spectra using 10 components for the SUBLIM reflectances, while as many as 23 components would be needed for the NATURAL data.

6.4.4 Discussion

By our analysis we have shown that spectral reflectances from different databases have different statistical properties. When fixing the amount of signal variance that should be accounted for, twice as many components

---

[11] ftp://ftp.eos.ncsu.edu/pub/spectra

Table 6.5: *Accumulated energy* $E_a(\tilde{P})$ *of the different databases. The entries corresponding to design choices of* $E_{req} = 0.90$ *and* $E_{req} = 0.99$ *are underlined, and reported in Table 6.6.*

$\tilde{P}$	MUNSELL	NATURAL	OBJECT	PIGMENT	SUBLIM
1	0.4783	0.4235	0.4833	0.4344	0.4005
2	0.6955	0.5836	0.6710	0.6714	0.6724
3	0.8288	0.7265	0.7841	0.7944	0.8491
4	0.8763	0.7953	0.8345	0.8583	0.9108
5	0.9094	0.8363	0.8750	0.8992	0.9426
6	0.9282	0.8695	0.9067	0.9242	0.9626
7	0.9446	0.8999	0.9273	0.9449	0.9737
8	0.9554	0.9204	0.9462	0.9593	0.9825
9	0.9652	0.9344	0.9581	0.9701	0.9871
10	0.9718	0.9448	0.9674	0.9788	0.9907
11	0.9767	0.9533	0.9749	0.9848	0.9926
12	0.9801	0.9596	0.9795	0.9888	0.9943
13	0.9829	0.9655	0.9836	0.9917	0.9951
14	0.9854	0.9701	0.9875	0.9942	0.9960
15	0.9871	0.9738	0.9903	0.9955	0.9967
16	0.9883	0.9766	0.9925	0.9963	0.9972
17	0.9896	0.9791	0.9941	0.9971	0.9975
18	0.9908	0.9814	0.9954	0.9977	0.9978
19	0.9917	0.9835	0.9962	0.9981	0.9981
20	0.9925	0.9854	0.9970	0.9984	0.9984
21	0.9933	0.9872	0.9976	0.9987	0.9986
22	0.9941	0.9889	0.9981	0.9989	0.9988
23	0.9949	0.9904	0.9985	0.9991	0.9990
24	0.9956	0.9918	0.9989	0.9993	0.9991
25	0.9963	0.9932	0.9992	0.9994	0.9993
26	0.9970	0.9945	0.9994	0.9996	0.9994
27	0.9977	0.9957	0.9996	0.9997	0.9996
28	0.9983	0.9969	0.9997	0.9998	0.9997
29	0.9989	0.9981	0.9998	0.9999	0.9998
30	0.9995	0.9991	0.9999	0.9999	0.9999
31	1.0000	1.0000	1.0000	1.0000	1.0000

Table 6.6: *Effective dimension D_e for the different databases for a choice of required accumulated energy of $E_{req} = 0.90$ and $E_{req} = 0.99$.*

E_{req}	MUNSELL	NATURAL	OBJECT	PIGMENT	SUBLIM
0.90	5	8	6	6	4
0.99	18	23	15	13	10

are needed to encode a spectrum from the NATURAL database than for the SUBLIM data. These results have quite important practical consequences when designing a multispectral image acquisition system.

The goal of this study is to draw conclusions on the importance of the data set being adapted to the application. Is it important to use oil painting

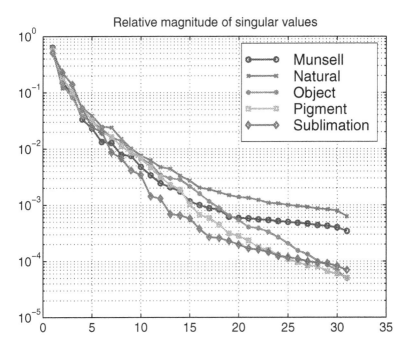

Figure 6.16: *Comparison of singular values of the 5 different databases in logarithmic scale. The steeper the curves decrease, the more the energy is concentrated in the first singular vectors.*

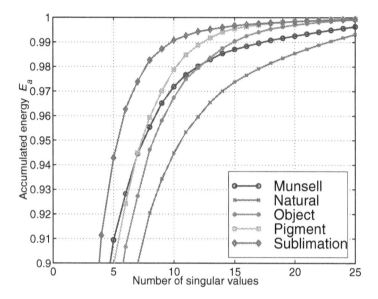

Figure 6.17: *Comparison of the accumulated energy* $E_a(\tilde{P})$ *of the different databases. For example if* $E_{req} = 99\%$ *of the signal energy should be preserved, 23 singular vectors should be used for the* NATURAL *data, while only 10 are requested for the* SUBLIM *data.*

reflectances when designing a multispectral image acquisition system for paintings, or could a standard set of Munsell reflectances equally well be used? Unfortunately, time did not allow us to further elaborate this study. Several other measures could have been applied to these reflectance data sets, such as the similarity measure proposed by Sharma *et al.* (1998), the Vora-measure (Vora and Trussell, 1993), a measure of how good a set of reflectance curves is approximated with a set of basis function proposed by Mahy *et al.* (1994b), and the goodness-fitting-coefficients proposed by Romero *et al.* (1997) and García-Beltrán *et al.* (1998).

6.5 Choice of the analysis filters

The quality of the spectral reflectance reconstruction depends not only on the reconstruction operator, but also heavily on the spectral characteristics of the acquisition system: illuminant, camera and filters. The design of optimal filters given an optimization criterion has been proposed by several authors, as for example in (Vora *et al.*, 1993, Vora and Trussell, 1997, Vrhel and Trussell, 1994, Vrhel *et al.*, 1995, Sharma and Trussell, 1996b, Lenz *et al.*, 1995; 1996a, Wang *et al.*, 1997). A drawback with such methods is the cost and difficulty involved in the production of the optimized filters. Another approach encountered in most existing multispectral scanner systems is to use a set of heuristically chosen color filters, which are typically equi-spaced over the visible spectrum (Burns, 1997, Keusen and Praefcke, 1995, Keusen, 1996, König and Praefcke, 1998b, Martinez *et al.*, 1993, Abrardo *et al.*, 1996). For example, the VASARI scanner implemented at the National Gallery in London uses seven broad-band nearly-Gaussian filters covering the visible spectrum (Martinez *et al.*, 1993). Although promising results are reported using such systems, the choice of filters seems to remain rather heuristic and likely sub-optimal.

An intermediate solution can be used where the camera filters are chosen from a set of readily available filters (Vora *et al.*, 1993, Maître *et al.*, 1996, Vora and Trussell, 1997). This choice is optimized, taking into account the statistical spectral properties of the objects that are to be imaged, as well as the spectral transmittances of the filters, the spectral characteristics of the camera, and the spectral radiance of the illuminant, *cf.* Section 6.2.2. The main idea is to choose the filters so that, when multiplied with the illuminant and camera characteristics, they span the same vector space as the reflectances that are to be acquired in a particular application, as suggested *e.g.* by Chang *et al.* (1989), Schmitt *et al.* (1990), Vora and Trussell (1993), Mahy *et al.* (1994b). Different schemes for selecting the filters are presented in the following section.

6.5.1 Filter selection methods

In this section we review different methods to select a subset of $\tilde{K}$ filters out of a set of K available filters. We suppose known the spectral transmittances $\phi_k(\lambda)$, $k = 1 \ldots K$, of the filters as well as the spectral sensitivity $\omega(\lambda)$ of the camera. After having mixed these functions together, we represent the filters (or more precisely the associated camera channel sensitivities) by the vectors $\mathbf{y}_k$,

$$\mathbf{y}_k = \alpha_k [\phi_k(\lambda_1)\omega(\lambda_1), \ \phi_k(\lambda_2)\omega(\lambda_2), \ \ldots, \ \phi_k(\lambda_N)\omega(\lambda_N)]^t,$$
$$k = 1\ldots K. \tag{6.34}$$

The normalization factors α_k *may* be chosen such that $\|\mathbf{y}_k\| = 1$.

The goal is to select, among a set of K available color filters, a subset of $\tilde{K}$ filters being well suited for our spectral reconstruction.

6.5.1.1   Equi-spacing of filter central wavelengths

A simple strategy is to choose a set of filters where the central wavelengths are relatively equally spaced throughout the visible spectrum (Martinez *et al.*, 1993, Keusen and Praefcke, 1995, Keusen, 1996, König and Praefcke, 1998b, Abrardo *et al.*, 1996, Burns, 1997).

6.5.1.2   Exhaustive search

It is clear that an optimal solution would emerge from a combinatorial approach, where all possible filter combinations are evaluated. The complexity of such an approach could be prohibitive, since it requires the evaluation of

$$n_c = \binom{K}{\tilde{K}} = \frac{K!}{\tilde{K}!(K - \tilde{K})!} \tag{6.35}$$

filter combinations. For a small number of filters, this method may be applicable, see *e.g.* Yokoyama *et al.* (1997) who evaluates the $n_c = 80730$ combinations needed for a selection of $\tilde{K} = 5$ filters from a set of $K = 27$, or Vora *et al.* (1993), Vora and Trussell (1997) who selects $\tilde{K} = 3$ filters from a set of $K = 100$ Wratten filters, requiring $n_c = 1.6 \times 10^5$ filter combinations. However, when the number of available filters as well as the number of filters to be chosen increase, the complexity grows considerably, as shown in Figure 6.18. For the example presented by Maître *et al.* (1996), where $K = 37$ and $\tilde{K} = 12$, the number of filter combinations to be evaluated attains $n_c = 1.8 \times 10^9$, giving a computation time in the order of days on a 100 MFLOP/s computer.

To reduce the computational cost, several constructive approaches are proposed in the following sections, taking into account the spectral properties of the available filters, the acquisition system, as well as the statistical spectral properties of the surfaces that are to be imaged.

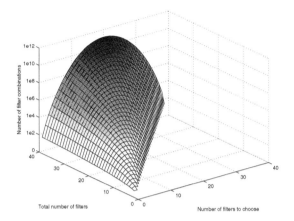

Figure 6.18: *Illustration of the computational complexity involved when comparing all possible filter combinations from a set of real filters. Note the logarithmic ordinate axis.*

6.5.1.3 Progressive optimal filters

As seen in Section 6.4.1, a set of reflectances may be characterized by its *characteristic reflectances*, that is, an orthogonal basis of the vector space spanned by the reflectances, sorted in order of decreasing importance. The idea of using linear combinations of the characteristic reflectances as filter transmittances was introduced by Chang *et al.* (1989). Mahy *et al.* (1994b) introduces the notion of *progressive optimal filters*, where the kth progressive optimal filter is defined as a linear combination of the first k characteristic reflectances, having only positive values. The exact manner in which this linear combination is determined is not described in the paper. They do not consider how to realize in practice the progressive optimal filters.

We propose here to select the kth filter from the set of available filters as the one that shows the highest correlation to the progressive optimal filter.

6.5.1.4 Maximizing filter orthogonality

The idea of this method is to choose a set of filters whose non-normalized [12] vectors $\mathbf{y}_k$ have a maximized orthogonality. The algorithm consists of the following steps.

1. As the first filter $\mathbf{b}_1 = \mathbf{y}_{k_1}$ we choose the one of maximal norm: $\|\mathbf{y}_{k_1}\| \geq \|\mathbf{y}_k\|$, $k = 1 \ldots K$. That is, the filter that transfers most energy is chosen.

2. The second filter $\mathbf{b}_2 = \mathbf{y}_{k_2}$ is chosen among the other filters as maximizing its component orthogonal to $\mathbf{b}_1$:

$$k_2 = \arg \max_{\substack{k \\ 1 \leq k \leq K \\ k \neq k_1}} \left\| \mathbf{y}_k - \mathbf{b}_{n1}(\mathbf{b}_{n1}^t \mathbf{y}_k) \right\|, \tag{6.36}$$

where $\mathbf{b}_{n1}$ denote the normalized vector $\mathbf{b}_{n1} = \mathbf{b}_1/\|\mathbf{b}_1\|$.

3. We continue for the $(i+1)$th vector which is chosen if it maximizes its component normal to the space $R([\mathbf{b}_1, \mathbf{b}_2, \ldots, \mathbf{b}_i])$, spanned by the set of vectors $\{\mathbf{b}_1, \mathbf{b}_2, \ldots, \mathbf{b}_i\}$.

   We denote the orthonormal basis of the space $R([\mathbf{b}_1, \mathbf{b}_2, \ldots, \mathbf{b}_i])$ as $\mathbf{B}_n^{(i)}$, which is constructed iteratively as

$$\mathbf{B}_n^{(i)} = \left[\mathbf{B}_n^{(i-1)} \, \mathbf{b}_{ni} \right], \tag{6.37}$$

where $\mathbf{B}_n^{(1)} = \mathbf{b}_{n1}$, and

$$\mathbf{b}_{ni} = \frac{\mathbf{b}_i - \mathbf{B}_n^{(i-1)}(\mathbf{B}_n^{(i-1)t} \mathbf{b}_i)}{\|\mathbf{b}_i - \mathbf{B}_n^{(i-1)}(\mathbf{B}_n^{(i-1)t} \mathbf{b}_i)\|}. \tag{6.38}$$

We then choose the $(i+1)$th basis vector $\mathbf{b}_{i+1} = \mathbf{y}_{k_{i+1}}$ for the $k = k_{i+1}$ that maximizes the following expression:

$$k_{i+1} = \arg \max_{\substack{k \\ 1 \leq k \leq K \\ k \notin \{k_1, k_2, \ldots, k_i\}}} \left\| \mathbf{y}_k - \mathbf{B}_n^{(i)}(\mathbf{B}_n^{(i)t} \mathbf{y}_k) \right\| \tag{6.39}$$

---

[12] It would also be possible to apply this method to the normalized vectors $\mathbf{y}_k$. This would typically yield a selection of filters with high degree of orthogonality, but lower transmission factors. For the image acquisition this can be accounted for to a certain extent by increasing the exposure time and/or aperture width, but for practical applications, it is clear that higher transmission factors are preferrable.

6.5.1.5   Maximizing orthogonality in characteristic reflectance vector space

This method[13] is more physically related to the problem which we have to solve. The central idea of the method is to select filters that have a high degree of orthogonality *after projection* into the vector space $R(\mathbf{U}^{(r)})$ spanned by the r most significant characteristic reflectances $\mathbf{u}_i, i = 1 \ldots r$, calculated by PCA/SVD analysis of a set $\mathbf{R}$ of sample reflectances. The matrix

$$\mathbf{U}^{(r)} = [\mathbf{u}_1 \mathbf{u}_2 \ldots \mathbf{u}_r], \quad r \leq \operatorname{rank}(\mathbf{R}) \tag{6.40}$$

represents thus the orthonormal basis of the vector space $R(\mathbf{U}^{(r)})$, *cf.* Eq. 6.24.

The projection of the kth filter on the jth characteristic reflectance vector is $\mathbf{u}_j^t \mathbf{y}_k$ and its projection in $R(\mathbf{U}^{(r)})$ is denoted as the $r \times 1$ coordinate vector $\mathbf{g}_k = \mathbf{U}^{(r)t} \mathbf{y}_k$. Note that $\mathbf{g}_k$ corresponds to the camera responses through the kth filter to a set of characteristic reflectances $\mathbf{U}^{(r)}$.

1. Considering the set of projections $\mathbf{g}_k, k = 1 \ldots K$, we choose as the first basis vector $\mathbf{y}_{k_1}$ the one which transfers most energy from the r most significant characteristic reflectances:

$$k_1 = \arg \max_{\substack{k \\ 1 < k < K}} \|\mathbf{g}_k\| \tag{6.41}$$

of maximal norm as the first basis vector $\mathbf{b}_1 = \mathbf{y}_{k_1}$. That is, the filter that transfers most energy from the characteristic reflectances is chosen.

2. The second filter $\mathbf{y}_{k_2}$ is then the filter whose projection onto $R(\mathbf{U}^{(r)})$ has a maximal component orthogonal to $\mathbf{g}_{k_1}$:

$$k_2 = \arg \max_{\substack{k \\ 1 < k < K \\ k \neq k_1}} \left\| \mathbf{g}_k - \mathbf{g}_{nk_1} \left(\mathbf{g}_{nk_1}^t \mathbf{g}_k \right) \right\|, \tag{6.42}$$

where $\mathbf{g}_{nk_1} = \mathbf{g}_{k_1} / \|\mathbf{g}_{k_1}\|$.

3. Let $\mathbf{G}^{(i)} = [\mathbf{g}_{k_1}, \mathbf{g}_{k_2}, \ldots, \mathbf{g}_{k_i}]$ denote the projections of the i first selected filters in $R(\mathbf{U}^{(r)})$. The filter $\mathbf{y}_{k_{i+1}}$ is then chosen such that

---

[13] The presented method is slightly modified compared to what we presented in (Maître *et al.*, 1996).

its projection $\mathbf{g}_{k_{i+1}} = \mathbf{U}^{(r)t} \mathbf{y}_{k_{i+1}}$ has the largest component orthogonal to $R(\mathbf{G}^{(i)})$.

The orthonormal basis of the space $R(\mathbf{G}^{(i)})$ spanned by the selected filters projected onto the characteristic reflectance space is denoted $\mathbf{G}_n^{(i)}$. It could be determined easily by a SVD applied to $\mathbf{G}^{(i)}$. However, this would imply a complete recalculation of the basis for each iteration. We propose to determine it in an iterative manner as follows. The first component is determined simply in step 1 by $\mathbf{G}_n^{(1)} = \mathbf{g}_{nk_1}$. For the ith iteration step, $\mathbf{G}_n^{(i)} = [\mathbf{G}_n^{(i-1)} \, \mathbf{g}_{ni}]$, where

$$\mathbf{g}_{ni} = \frac{\mathbf{g}_i - \mathbf{G}_n^{(i-1)}(\mathbf{G}_n^{(i-1)t} \mathbf{g}_i)}{\left\| \mathbf{g}_i - \mathbf{G}_n^{(i-1)}(\mathbf{G}_n^{(i-1)t} \mathbf{g}_i) \right\|} \qquad (6.43)$$

We then choose the $(i+1)$th basis vector $\mathbf{y}_{k_{i+1}}$ for the $k = k_{i+1}$ that maximizes the following expression:

$$k_{i+1} = \arg \max_{\substack{k \\ 1 \leq k \leq K \\ k \notin \{k_1, k_2, \ldots, k_i\}}} \left\| \mathbf{g}_k - \mathbf{G}_n^{(i)} \left(\mathbf{G}_n^{(i)t} \mathbf{g}_k \right) \right\| \qquad (6.44)$$

By this algorithm, given the choice of the number of characteristic vectors r that are taken into account, we can choose a set of $\tilde{K}$ filters, having spectral transmittances of $\phi_k(\lambda)$, $k = k_1, k_2, \ldots, k_{\tilde{K}}$.

6.5.2 Discussion

The presented methods present several advantages and disadvantages. We resume some of their key features in Table 6.7. The heuristic selection of equi-spaced filters is simple and intuitive, but clearly not optimized. An exhaustive combination of all filter combination can fulfill any optimization criterion, but if a large number of filters are to be considered, it is too slow. The progressive optimal filter method takes into account information about the spectral characteristics of the filters, camera, illuminant and the reflectances that will be imaged. However, the method is clearly sub-optimal since each filter is chosen sequentially considering only the corresponding *progressive optimal filter* regardless of the filters already chosen.

   In the last two methods the vector sub-space spanned by the $(i-1)$ chosen filters is considered when selecting the ith filter. The first of them maximizes the orthogonality, that is, the independence of the camera channels regardless of the imaged reflectances. In the last method, we maximize the orthogonality when applied to a set of characteristic reflectances. By doing this we adapt our filter selection to one specific application, and we achieve thus better results than with a general selection for this application. We have obtained very promising results by using this last method proposed in Section 6.5.1.5, these results are presented in Section 6.6.

Table 6.7: *Comparison of key properties of the proposed filter selection methods.*

Selection method	Considered information	Comput. complex.	Optimization criterion	Scheme
Equi-spacing (6.5.1.1)	None	Immediate	None	Heuristic
Exhaustive search (6.5.1.2)	*a posteriori*	Very high	Any	Combinatorial
Progressive optimal (6.5.1.3)	Filter, camera, illuminant, reflectances	Very low	Correlation to optimal filters	Each filter chosen individually. Clearly sub-optimal
Max orthogonality (6.5.1.4)	Filter, camera, illuminant	Low	Filter/channel orthogonality	Sub-optimal. Constructive (each filter chosen considering already chosen filters)
Orthog. in charac. reflectance space (6.5.1.5)	Filter, camera, illuminant, reflectances	Low	"Camera response" orthogonality	Sub-optimal. Constructive

Unfortunately, time did not allow us to perform a comprehensive testing and comparison of the described filter selection methods. One interesting experiment would be to first select a reduced number of filters, say 15, using our approach, and then compare the performance of the different methods in selecting a smaller subset, say 5, of these filters. With such a rather small number of filters, an exhaustive search would be realizable, thus we could compare all the methods to this optimal selection.

6.6   Evaluation of the acquisition system

We resume in Figure 6.19 the complete chain of a multi-channel image acquisition system with the final spectral reconstruction step.

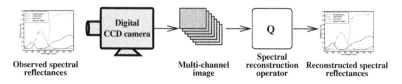

Observed spectral reflectances   Multi-channel image   Spectral reconstruction operator   Reconstructed spectral reflectances

Figure 6.19: *The complete chain of multi-channel image acquisition system with the final spectral reconstruction step.*

We have performed a simulation to evaluate the complete multispectral image acquisition system. We used D65 as the scanning illuminant, the Eikonix CCD camera spectral characteristics, filters chosen from a set of 37 Wratten, Hoffman, and Schott filters, and the spectral reflectances of a color chart of 64 pure pigments used in oil painting. The resulting spectral sensitivities of the camera channels, following the selection method described in Section 6.5.1.5, are shown in Figure 6.20(b) for the case of a selection of seven filters with transmittances shown in Figure 6.20(a). We note that, as expected, the peak sensitivities of the camera channels are distributed over the entire wavelength interval; however, they are not equally spaced.

To evaluate the quality of the multispectral acquisition system with a given set of filters, it makes sense to compare the original reflectances with the ones estimated from the camera responses. In Figure 6.21(a) we show some examples of spectral reflectance reconstruction, along with the corresponding RMS spectral reconstruction errors, that is, the Euclidian distance between original and reconstructed spectral reflectances.[14] We see that smooth reflectances such as the "Emerald Green" is reconstructed very precisely, while the the reconstruction of the "Mercuric Iodide" pigment which has a very steep transition near 590 nm, is much less accurate. When considering, for all the oil pigment reflectances, the maximal and minimal signed reconstruction errors for each wavelength, we present the

---

[14]The spectral reflectances having values between 0 and 1, the extreme example of a perfect white being reconstructed as a perfect black, would give a "maximal" RMS error of 1. A mean RMS error of 0.0357 as seen in the first column of Table 6.8 can then be roughly interpreted as a proportion of 3.57% of the maximal error.

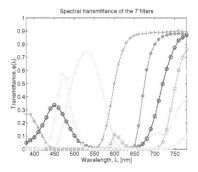 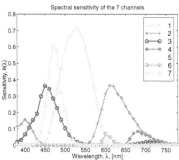

(a) The spectral transmittances of the seven filters

(b) The corresponding spectral sensitivity of the seven camera channels including the spectral characteristics of the Eikonix camera, the spectral radiance of the illuminant D65, and the spectral transmittance of the filters

Figure 6.20: *An acquisition system using seven filters chosen according to the method described in Section 6.5.1.5. (The numbers in (b) denote the sequence in which the filters are chosen.)*

spectral reconstruction error band in Figure 6.21(b). By comparing this diagram with Figure 6.20(b) we see that low maximal absolute errors occur approximately for the peak wavelengths of the channel sensitivites.

To quantify the quality of the multispectral image acquisition system we have chosen to use the mean and maximum RMS spectral reconstruction errors as a quality measure. This measure presents the advantage to be simple and general. In Table 6.8, the RMS spectral reconstruction errors using different number of filters are reported. As expected, we see that the mean reconstruction error decreases when an increasing number of filters are used. The maximum error shows some exceptions to this, however it follows the same decreasing trend.

Other quality measures could also have been used (Neugebauer, 1956, Vora and Trussell, 1993, Tajima, 1996, Sharma and Trussell, 1996a; 1997b). Depending on the intent, these may be based on colorimetric or spectral properties, on mean or maximal errors in a data set, or alternatively on critical samples for which the reconstruction quality is particularly impor-

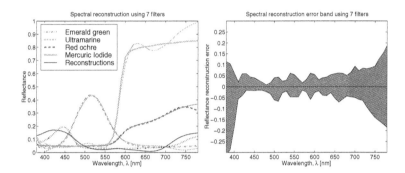

(a) Four examples of the reconstruction of spectral reflectances from the camera responses using the reconstruction operator Q_1. The RMS error for the Emerald green pigment is: 0.0122, Ultramarine: 0.0224, Red ochre: 0.0043, and Mercuric Iodide: 0.0322

(b) Spectral reconstruction error band over all the reflectances of the oil pigment chart. Low maximal errors occur approximately for peak wavelengths of the channel sensitivites, cf. Figure 6.20(b)

Figure 6.21: *Evaluation of the acquisition system using seven filters chosen according to the method described in Section 6.5.1.5.*

tant for a specific application. In the next section we will evaluate the colorimetric quality of the system, that is, to which degree the color of a surface as it would appear under a given illuminant can be estimated from the acquired multispectral image.

Table 6.8: *Comparison of the RMS spectral reconstruction error for varying number of filters using the reconstruction operator Q_1 (cf. Section 6.3).*

Number of filters	3	4	5	6	7
Mean RMS error	0.0357	0.0239	0.0178	0.0132	0.0111
Max RMS error	0.0879	0.0677	0.0538	0.0493	0.0616
	8	9	10	11	12
	0.0087	0.0057	0.0056	0.0036	0.0030
	0.0323	0.0174	0.0184	0.0122	0.0105

6.7 Multimedia application: Illuminant simulation

In the previous sections we have presented different aspects of the acquisition of a multispectral image. This multispectral image may be used for many purposes: object recognition, color constancy, high-quality reproduction, etc. We present here a particular application, which is the simulation of the original scene as it would have appeared when viewed under different illuminants. Applied to fine arts paintings, museological objects, jewelry, textiles, etc., such simulations displayed on a color calibrated computer monitor could be of particular interest in

- a high-end multimedia application for the open market, the user himself choosing his preferred light source, or

- a computer aided tool for specialists, for example a curator having to decide the appropriate light sources for an art exhibition.

It is well known that the appearance of an object or a scene may change considerably when the illuminant changes, due to physical and psychophysical effects. These effects are taken into account in most color appearance models in a somewhat heuristic manner. However, such models cannot predict correctly changes for arbitrary illuminants, one important reason for this being metamerism. To make quantitative predictions about the physical phenomena involved when the illuminant is changed, a complete spectral description of the illuminants and the scene reflectances is required.

We will here present two methods for the simulation of objects viewed under different illuminants. First, a classical method based on the CIELAB space is described in Section 6.7.1. Then, we describe in Section 6.7.2 the method applying multispectral imaging techniques. In Section 6.7.3 we compare the two methods using ΔE_{94}^* under the simulated illuminant as an error measure.

6.7.1 Illuminant simulation using CIELAB space

It is found by several studies (Lo *et al.*, 1996, Braun and Fairchild, 1997) that the CIELAB space (CIE 15.2, 1986) performs well in simulating a change in illuminant, and that it can be compared to more complicated color appearance models such as RLAB (Fairchild and Berns, 1993) or the Hunt model (Hunt, 1995). It is clear, however, that CIELAB does not make an attempt to take into account parameters such as ambient light, surround, etc.

To evaluate the ability of CIELAB space to account for changes in viewing illuminant, we first define an ideal colorimetric image capture device having its spectral sensitivities equal to the color matching functions of the CIE XYZ-1931 standard observer (cf. Figure 2.9 on page 25), and for which we use D65 as illuminant. The three channels of this ideal camera provide us directly with the exact tristimulus values of the surface imaged in each pixel,

$$[X_{D65}\ Y_{D65}\ Z_{D65}]^t = \mathbf{A}^t \mathbf{L}_{D65}\ \mathbf{r}, \qquad (6.45)$$

where $\mathbf{A} = [\bar{\mathbf{x}}\ \bar{\mathbf{y}}\ \bar{\mathbf{z}}]$ represents the color matching functions, and $\mathbf{L}_{D65}$ is a diagonal matrix containing the D65 spectral radiance.

The key point in the way CIELAB treats the illuminant is that when converting from XYZ to CIELAB, the XYZ values are taken *relative* to the XYZ values of the illuminant. Thus,

$$[L_{D65}^*, a_{D65}^*, b_{D65}^*] = g\left(\frac{X_{D65}}{X_{W,D65}}, \frac{Y_{D65}}{Y_{W,D65}}, \frac{Z_{D65}}{Z_{W,D65}}\right), \qquad (6.46)$$

the function $g(\cdot)$ being defined by the well-known functions given in Section 2.4.8.1, and $[X_{W,D65}, Y_{W,D65}, Z_{W,D65}]$ being the tristimulus values of a perfect diffuser under D65 lighting. Since we assume an ideal image capture device, these CIELAB values are colorimetrically exact for illuminant D65.

When using CIELAB as a color appearance model, we assume that the CIELAB values of a given surface color are constant and independent of illuminant changes. The estimation of the CIELAB values of this color under a simulated illuminant L_{sim} are thus given by

$$[\hat{L}_{\text{sim}}^*, \hat{a}_{\text{sim}}^*, \hat{b}_{\text{sim}}^*] = [L_{D65}^*, a_{D65}^*, b_{D65}^*]. \qquad (6.47)$$

By applying the inverse transformation $g^{-1}(\cdot)$ to Equation 6.47, we obtain the following relation

$$\left[\frac{\hat{X}_{\text{sim}}}{X_{W,\text{sim}}}, \frac{\hat{Y}_{\text{sim}}}{Y_{W,\text{sim}}}, \frac{\hat{Z}_{\text{sim}}}{Z_{W,\text{sim}}}\right] = \left[\frac{X_{D65}}{X_{W,D65}}, \frac{Y_{D65}}{Y_{W,D65}}, \frac{Z_{D65}}{Z_{W,D65}}\right], \qquad (6.48)$$

where $[\hat{X}_{\text{sim}}, \hat{Y}_{\text{sim}}, \hat{Z}_{\text{sim}}]$ and $[X_{W,\text{sim}}, Y_{W,\text{sim}}, Z_{W,\text{sim}}]$ are the tristimulus values of the surface and of the perfect diffuser, respectively, under the illuminant L_{sim}. We see from Equation 6.48 that the CIELAB space takes into account the effects of chromatic adaptation by applying a von Kries-like (von Kries, 1902) transform in the XYZ space.

6.7.2 Illuminant simulation using multispectral images

We now consider the simulation of spectral changes in lighting if multispectral images are available. Such images may be acquired as described in the previous sections or by other means, the essential being that they contain, in each pixel, information from which the spectral reflectance imaged on it can be reconstructed.

To simulate the scene as it would have appeared when lit by a given illuminant L_{sim}, the multispectral image provides us with a straightforward approach: the reconstructed spectra $\tilde{r}$ in each pixel is first reconstructed from its multispectral coordinates c_K using Equation 6.16, $\tilde{r} = Qc_K$. We then calculate colorimetrically the estimated XYZ tristimulus values of the surface imaged in this pixel and lit by illuminant L_{sim} as in Equation 6.45:

$$[\tilde{X}_{sim}, \tilde{Y}_{sim}, \tilde{Z}_{sim}]^t = A^t L_{sim} \tilde{r}, \qquad (6.49)$$

where L_{sim} is the diagonal matrix corresponding to the spectral radiance of the simulated illuminant. These values are then used to estimate the CIELAB values under this particular illuminant:

$$[\tilde{L}^*_{sim}, \tilde{a}^*_{sim}, \tilde{b}^*_{sim}] = g\left(\frac{\tilde{X}_{sim}}{X_{W,sim}}, \frac{\tilde{Y}_{sim}}{Y_{W,sim}}, \frac{\tilde{Z}_{sim}}{Z_{W,sim}}\right) \qquad (6.50)$$

6.7.3 Evaluation of the two illuminant simulation methods

When evaluating the ability of different methods to take into account a change in illuminant, psychophysical tests using real observers should be applied (Lo *et al.*, 1996, Braun and Fairchild, 1997). However, a numerical criterion for this evaluation may also be of great interest because of its simplicity and rapidity. For example we may perform an analysis based on the CIE ΔE^*_{94} (McDonald and Smith, 1995). For a simulated illuminant L_{sim}, the exact CIELAB values under this illuminant are calculated as follows

$$[X_{sim}, Y_{sim}, Z_{sim}]^t = A^t L_{sim} r, \qquad (6.51)$$

$$[L^*_{\text{sim}}, a^*_{\text{sim}}, b^*_{\text{sim}}]^t = g\left(\frac{X_{\text{sim}}}{X_{W,\text{sim}}}, \frac{Y_{\text{sim}}}{Y_{W,\text{sim}}}, \frac{Z_{\text{sim}}}{Z_{W,\text{sim}}}\right). \qquad (6.52)$$

These values are then compared to the estimated values by the CIELAB model, $[\hat{L}^*_{\text{sim}}, \hat{a}^*_{\text{sim}}, \hat{b}^*_{\text{sim}}]$ (cf. Equation 6.47), and to those estimated by the multispectral image approach $[\tilde{L}^*_{\text{sim}}, \tilde{a}^*_{\text{sim}}, \tilde{b}^*_{\text{sim}}]$ (cf. Equation 6.50).

We have performed an analysis of the illuminant-simulation quality for the multispectral image approach with 5, 7 and 10 channels, and for the CIELAB space as a color appearance model with D65 as starting reference. These four methods are evaluated using five illuminants: the CIE daylight illuminants D65 and D50, the CIE standard illuminant A (representative of a typical tungsten lighting with a color temperature of 2856K), a normal fluorescent lamp F2, and a low-pressure sodium lamp (LPS) widely used in street lighting (see Figure 6.22). The spectral reflectances used for evaluation are those of the 64 oil pigments previously introduced in Section 6.5.

The results in terms of mean and maximal ΔE^*_{94} errors are listed in Table 6.9, the error histograms are given in Figure 6.23, and a graphical representation of the results for the case of a seven-channel acquisition system is given in Figure 6.24. The obtained results are found to be comparable to those obtained in previous research, e.g. by Vrhel and Trussell (1992; 1994).

Table 6.9: *Mean and maximal ΔE^*_{94} errors obtained for the simulations of five illuminants with four different methods: CIELAB space used as a color appearance model and the three multispectral approaches using 5, 7 and 10 filters, respectively.*

Simulated illuminant	CIELAB		Multisp. (5)		Multisp. (7)		Multisp. (10)	
	Mean	Max	Mean	Max	Mean	Max	Mean	Max
D65	0.00	0.00	1.58	10.53	0.56	2.45	0.14	0.53
A	4.94	11.33	1.92	15.51	0.54	3.90	0.14	0.67
F2	3.63	7.67	2.15	14.31	0.71	3.79	0.31	2.00
D50	1.56	4.27	1.71	12.23	0.56	2.87	0.14	0.54
LPS	20.10	52.68	1.40	10.06	1.37	11.61	1.01	7.82

We note that at the evident exception of D65 which serves as reference for the CIELAB model, the multispectral approach performs generally significantly better than the CIELAB model. For example for the illuminant A, the mean error is approximately ten times smaller using the multispectral approach with 7 filters than with the CIELAB model. The CIELAB

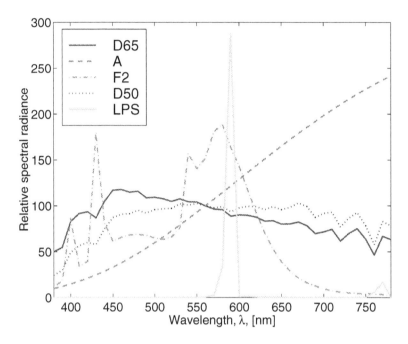

Figure 6.22: *Relative spectral radiances of the five illuminants used in the experiment.*

model performs reasonably well for the D50 case. This was expected since D65 and D50 have similar spectra, the change from D65 to D50 introducing only limited metameric problems. The D50 illuminant simulation using the CIELAB model is even better than the multispectral approach when using only 5 filters (Mean $\Delta E_{94}^* = 1.71$), the spectral reconstruction errors becoming greater than the errors induced by the CIELAB model (Mean $\Delta E_{94}^* = 1.56$). Almost complete failure, with a mean ΔE_{94}^* error of 20.10, is found for the CIELAB model in the case of low-pressure sodium (LPS) lamp. This was also expected, since its spectral power distribution consists almost entirely of two spectral lines at 589.0 and 589.6 nm (Hunt, 1991). We see also that if we only look at the maximal errors, the CIELAB method outperforms the 5-filter multispectral approach (except for the LPS lamp).

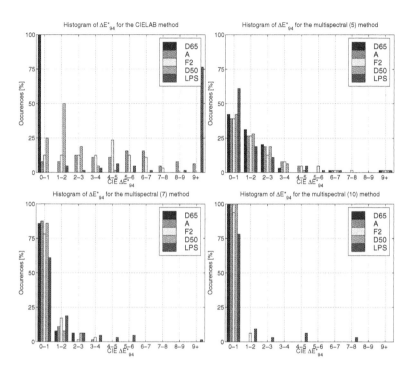

Figure 6.23: *Histograms of ΔE_{94}^* simulation errors for the CIELAB method and the multispectral methods using 5, 7, and 10 filters. The models' performance is compared to direct spectral calculation of CIELAB under the simulated illuminant.*

6.8   Conclusion

We have described several aspects concerning the design, application and setup of a system for the acquisition of multispectral images in which a set of chromatic filters are used with a CCD camera.

We have presented several approaches to the problem of spectral characterization of an electronic camera. The characterization is obtained by measuring a set of patches of known spectral reflectances and by inverting the resulting system of linear equations. In the presence of noise, this system inversion is not straightforward. We have shown that the choice of samples is of great importance for the quality of the characterization,

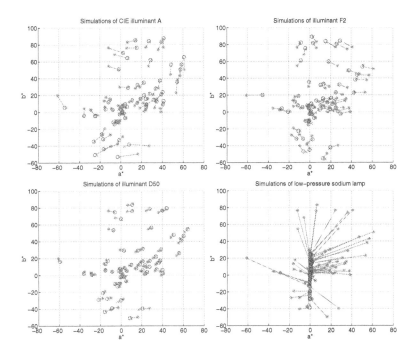

Figure 6.24: *Simulation results for a seven-channel acquisition system with the illuminants A, F2, D50 and a low-pressure sodium lamp. The models' performance is compared to direct spectral calculation of CIE-LAB under the simulated illuminant. The results are projected in the a*-b*-plane of the CIELAB space. The reference CIELAB values under the simulated illuminant are marked with circles (○), the values predicted by the CIELAB model by asterisks (*), and those predicted by the multispectral image approach by crosses (×). We note a clear superiority of the simulation obtained by using the multispectral image approach.*

and we have presented an algorithm for the selection of a reduced number of patches. Using this optimized selection method allowed us to get comparable performance using only 20 optimally selected color patches, as compared to using the spectral reflectances of the complete set of 1269 Munsell chips.

An efficient method was proposed for the estimation of the spectral reflectance of each pixel of the scene, from the camera responses using the set of filters. This reconstruction is optimized by taking into account the statistical spectral properties of the objects to be imaged, as well as the spectral characteristics of the camera, and the spectral radiance of the illuminant used for the acquisition.

We have performed a comparative statistical analysis of different sets of spectral reflectances, finding that the *effective dimensions* D_e of the different vector spaces spanned by the reflectances of the databases may be very different from one database to another. These results may give an indication on the number of acquisition filters that should be used for a given application, and also on how much the spectral information may be compressed.

We further proposed several methods for the selection of a set of filters that permits a good quality of the estimation of spectral reflectances from camera responses. The main idea is to choose the filters so that, when multiplied with the illuminant and camera characteristics, their orthogonality is maximized after projection in the characteristic reflectance vector space corresponding to a particular application.

We then proposed a methodology to evaluate the overall quality of a multispectral image acquisition system with a given set of filters. We were concerned in particular with the system's ability to produce accurate information about the spectral reflectances of the scene.

Finally, we presented an application in which multispectral imaging excels. In order to reveal the modifications in the color appearance of an object or a scene when the illuminant is changed, a colorimetric simulation can be of particular interest in multimedia applications, especially in the museum field. We have investigated two methods for such an illuminant simulation, a classical method using the CIELAB color space as a color appearance model, and a method using multispectral images. The multispectral image approach is found to perform very well, even when applied to illuminants that are particularly difficult to handle with conventional methods.

## Chapter 7

Multispectral image acquisition: Experimentation

IN THIS CHAPTER WE DESCRIBE THE EXPERIMENTAL SETUP OF A MUL-TISPECTRAL IMAGE ACQUISITION SYSTEM CONSISTING OF A PROFES-SIONAL MONOCHROME CCD CAMERA AND A TUNABLE FILTER IN WH-ICH THE SPECTRAL TRANSMITTANCE CAN BE CONTROLLED ELECTR-ONICALLY. WE HAVE PERFORMED A SPECTRAL CHARACTERIZATION OF THE ACQUISITION SYSTEM TAKING INTO ACCOUNT THE ACQUISI-TION NOISE. TO CONVERT THE CAMERA OUTPUT SIGNALS TO DEVICE-INDEPENDENT DATA, TWO MAIN APPROACHES ARE PROPOSED. ONE CONSISTS IN APPLYING AN EXTENDED VERSION OF THE COLORIMET-RIC SCANNER CHARACTERIZATION METHOD DESCRIBED PREVIOUSLY TO CONVERT FROM THE K CAMERA OUTPUTS TO A DEVICE-INDE-PENDENT COLOR SPACE SUCH AS CIEXYZ OR CIELAB. ANOTHER METHOD IS BASED ON THE SPECTRAL MODEL OF THE ACQUISITION SYSTEM. BY INVERTING THE MODEL USING A PRINCIPAL EIGENVEC-TOR APPROACH, WE ESTIMATE THE SPECTRAL REFLECTANCE OF EACH PIXEL OF THE IMAGED SURFACE.

7.1   Introduction

In the previous chapter, we have developed several algorithms concerning the acquisition of multispectral images, and their performances were evaluated by simulations. We will here examine how these algorithms perform in practice. An experimental multispectral camera was assembled using a *PCO SensiCam* monochrome CCD camera and a *CRI VariSpec* Liquid Crystal Tunable Filter (LCTF). This setup has permitted to test and validate several of the algorithms described in Chapter 6.

In Section 7.2 we describe the important features of our equipment, and in Section 7.3 we describe how he make sure that the camera response is linear with regards to the energy of the incident light in the entire scene. In Section 7.4 we perform a spectral characterization of the image acquisition system, and in Section 7.5 an experimental 17 channel multispectral image acquisition of the Macbeth chart is described. Finally, in Section 7.6 we examine how colorimetric and spectrophotometric information can be determined from the camera responses.

7.2   Equipment

In this section we describe the different components we have used in our experiments, and discuss briefly some of their important features.

7.2.1   CCD camera

The camera we used in our experiments is a SensiCam "Super-VGA" monochrome CCD camera from PCO Computer Optics GmbH.[1] It has a resolution of 1280×1024 pixels, a dynamic range of 12 bit, exposure times from 1 ms to 1000 s, and it operates at a 12.5 MHz readout frequency. The CCD is grade 0 (no defective pixels), and it is cooled to $-12\,^{\circ}C$ to reduce noise to a minimum. Its spectral sensitivity function as given by the manufacturer is reported in Figure 7.1. The camera is controlled from a PC via a PCI-board. It is delivered with a Software Developers Kit (SDK) which has enabled us to develop efficient image acquisition software corresponding to our needs.[2] By grouping several pixels within rows or columns with

---

[1] http://www.pco.de/

[2] This software have been developed in C and Java by Hans Brettel with the participation of Jerôme Neel.

a technique called *binning*, the sensitivity can be increased while reducing the resolution proportionally.

Although having a rather limited spatial resolution compared to our first camera (the Kodak Eikonix 1412 line-scan CCD camera described in Section 4.3) the SensiCam has the enormous advantage of being several orders of magnitude faster. This is of great importance since several acquisitions must be made for each scene in order to acquire a multispectral image. In our earlier experiments done with the Eikonix camera, this process was prohibitively slow. If a higher resolution is required, for example for fine-art paintings (Maître *et al.*, 1996), we can either upgrade to a higher resolution camera, once all the algorithms have been developed and tested, or apply mosaicing techniques as described in Section 4.5.1.

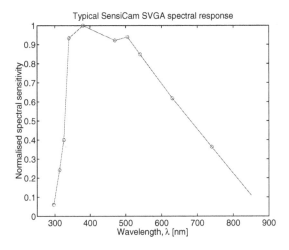

Figure 7.1: *Typical spectral sensitivity of PCO SensiCam CCD camera as given by the manufacturer (SensiCam Specification, 1997).*

7.2.2 Tunable filter

We recently acquired a Liquid Crystal Tunable Filter (LCTF), the *VariSpec* from Cambridge Research & Instrumentation (CRI), Inc.[3] This system is made of two units which provide us with two set-ups, one for narrow-band

---

[3]http://www.cri-inc.com/

filters, the other for wide-band filters. We will not go into details on the physics behind the functionality of such filters here (refer *e.g.* Chrien *et al.*, 1993, Kopp and Derks, 1997, Harding, 1997, Savin, 1998), but only mention that it consists of several consecutive layers of Lyot-type bi-refringent filters (Lyot, 1933, Wyszecki and Stiles, 1982, p.51), each layer containing linear parallel polarizers sandwiching a liquid crystal retarder element. Each layer is operating in a higher order than the previous ones, thus being able to select narrower bandpass characteristics of varying peak wavelengths. The peak wavelength can be controlled electronically from an external controller unit, or from a computer via a RS-232 interface, in the range [400 nm, 720 nm]. The average Full-Width-at-Half-Maximum (FWHM) bandwidth is approximately 5 nm or 30 nm for the narrow-band and wide-band set-up, respectively.

Compared to another type of tunable filters, the Acusto-Optical Tunable Filters (AOTF) (Chang, 1976), the LCTF technology offers a reasonably wide field of view ($\pm 7°$ from the normal axis) but, nevertheless, the limitation in field-of-view is a parameter that has to be treated with care for imaging applications.

The spectral transmittances of the filter when varying the peak wavelength in 10 nm steps from 400 to 720 nm was measured with the Ocean Optics Model SD100 spectrometer (Brettel *et al.*, 1997, Savin, 1998). See Figure 7.2. Several interesting conclusions can be drawn from these transmittance spectra:

- The transmittances have more or less a Gaussian-like shape, except for the wide-band set-ups at peak wavelengths > 650 nm.

- Clearly, there is an additional infrared filter present, since the filter spectral transmittances are cut at the red end of the spectrum.

- The FWHM is not constant; for the wide-band set-up, it varies from 15 to 80 nm.

- For the wide-band set-up and peak wavelengths ≤ 440 nm, the filters have an unwanted secondary peak at long wavelengths (see Figure 7.2(b)).

Even if the filter characteristics do not completely fulfill the manufacturer's promises, we consider it as a very valuable tool for multispectral imaging.

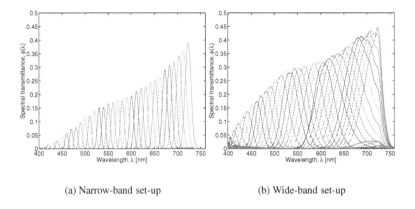

(a) Narrow-band set-up   (b) Wide-band set-up

Figure 7.2: *Spectral transmittances of the two different set-ups of the LTCF filter when varying the peak wavelength in 10 nm steps from 400 to 720 nm (Savin, 1998). Notice the unwanted secondary peaks in low wavelengths for the wide-band setup.*

7.2.3 Illumination

The importance of the illumination in image acquisition systems is often underestimated. Factors that need to be taken into consideration include the following:

- **Geometry.** The lamps should be placed to ensure a good spatial uniformity. A non-uniform lighting can however be corrected for, as described in Sections 4.4.1 and 7.3, so spatial uniformity is not necessarily a crucial requirement. If the camera is used for spectrophotometric or colorimetric measurement, a lighting/viewing geometry recommended by the CIE should be used, typically (45/0) in which the illuminant is placed at 45° off the normal axis, or (d/0) in which diffuse lighting is used, typically by means of an integrating sphere (CIE 15.2, 1986, Wyszecki and Stiles, 1982, p.155). For the acquisition of paintings, an important requirement is to avoid specular reflection on the painting surface, while for other types of objects (silverware, china, jewelry, etc.), specular reflections may be desired.

▓ **Power.** The lamps should have enough power to give a sufficient signal even through a narrow-band spectral filter. Low intensity can be compensated with long integration times, but that may pose additional problems, *e.g.* giving prohibitively slow acquisitions if a line-scan camera is used, or amplifying acquisition noise if the camera is not of high quality.

▓ **Spectral properties.** First, sufficient spectral power is needed in all parts of the visible spectrum. Secondly, if we seek to reconstruct the spectral reflectance of the scene, it is preferred that the spectral power distribution of the illuminant is as smooth as possible. If several lamps are used, it is also important that they have the same spectral power distribution, in order to guarantee the spatial homogeneity of the spectral power distribution.

▓ **Stability and repeatability.** The stability of the illumination is of utmost importance when a line-scan camera is used. But, if the camera is used to make precise, quantifiable acquisitions, the stability is also important for a CCD-matrix camera.

For our experiment we have used one 12V tungsten halogen lamp connected to a stabilized power supply. We used no diffuse reflectors to make sure that the spectral properties were spatially constant, and we also made sure there were no unwanted reflections by covering surrounding items with a black cloth.

7.2.4 Color chart

To characterize and evaluate our multispectral image acquisition system we chose to use the *Macbeth ColorChecker Color Rendition Chart* (McCamy *et al.*, 1976) because of its availability, its widespread use in color imaging (*e.g.* Farrell and Wandell, 1993, Burns, 1997, Finlayson *et al.*, 1998), and its spectral properties (Sec. 6.2.2.5). It is supposed to give a good representation of natural spectra. The chart and the measured spectral reflectances are shown in Figure 7.3.

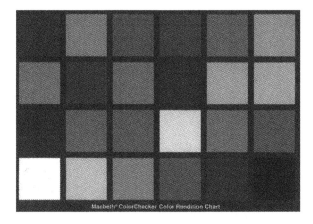

(a) Scanned color image

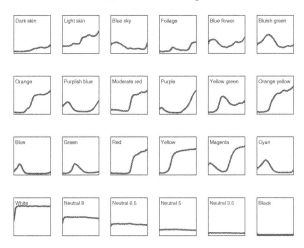

(b) Spectral reflectances measured with the Ocean Optics Model SD100 spectrometer. Displayed wavelength range is from 400 to 760 nm.

Figure 7.3: *The Macbeth ColorChecker Color Rendition Chart used in our experiments.*

7.3   Illumination and dark current compensation

For our analysis we need to make sure that the camera response is linear with regards to the energy of the incident light, *cf.* Section 3.2.2. We obtain this by correcting for the camera's dark current. Furthermore, we correct for the uneven distribution of the lighting, *cf.* Section 4.4.1.

The first step of our calibration is to measure the dark noise of the camera. To do so, we acquire a set of images where no light was entering the objective, *i.e.* with the lens cap on in a dark room, using varying integration times. We found that the dark noise was approximately $e_d = 60 \pm 5$ (on a scale of digital counts on 12 bit, giving values from 0 to $2^{12} - 1 = 4095$). This number shows almost no variation with the integration time. A slight augmentation was found only for integration times of several seconds, and we thus assume that the black noise is constant and independent of the acquisition parameters.

In the second step, we acquire an image $W(i, j)$ of a uniform diffuse surface. This provides us information about the spatial distribution of the illuminant. An image $I(i, j)$ of the Macbeth chart is then acquired, and we calculate a normalized image $I_n(i, j)$ as follows;

$$I_n(i, j) = k_I \frac{I(i, j) - e_d}{W(i, j) - e_d}, \tag{7.1}$$

the normalization factor k_I being chosen so that the pixel values of the normalized image are limited to a given maximal value. The images $W(i, j)$ and $I(i, j)$ are encoded as 12 bits per pixel, and the normalized image $I_n(i, j)$ on 16 bit per pixel and saved to a file for further use. Note that the normalization factor k_I is image-dependent. Having done this we verify the linearity of the image $I_n(i, j)$ by extracting the mean pixel values of the gray patches of the Macbeth chart, and comparing these to the reflectance factors of the patches, *cf.* Section 3.2.2. From Figure 7.4 we see that the data can be fitted reasonably well to a straight line, and we conclude thus that no further linearization is needed.

7.4   Spectral sensitivity estimation

The next step is to perform a spectral characterization of the image acquisition system, as described in Section 6.2. We recall from Equation 6.4

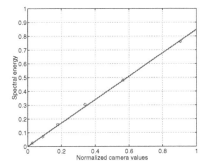

Figure 7.4: *Verification of linearity of the SensiCam CCD camera after normalization for black current and lighting distribution. The measured data fit reasonably well to a straight line, and no further linearization is needed.*

that the acquisition process can be modeled as[4]

$$c = \mathbf{r}^t \boldsymbol{\omega} + \epsilon, \tag{7.2}$$

where c is the camera response for an acquisition system with a spectral sensitivity of $\boldsymbol{\omega}$, to a surface with a spectral reflectance $\mathbf{r}$. The acquisition noise is denoted ϵ. The spectral sensitivity that we seek to estimate includes thus the camera and the illuminant.

7.4.1 Preliminary experiment

In a preliminary experiment we assumed that the acquisition system could be characterized *i)* simply using the spectral sensitivity provided by the manufacturer as given in Figure 7.1, and *ii)* supposing the halogen lamp spectral power distribution is equivalent to the illuminant A. We used then the linear model of Equation 7.2 with 5nm sampling intervals from 400 to 700 nm, to predict the camera output. In Figure 7.5 we compare the predicted camera responses to the experimentally observed responses for all the patches of the Macbeth chart. We have normalized the responses to a maximum of one for easy comparison. We see that there is a quite good fit for the grey patches (19-24), while for the other colors the differences

---

[4]Since we here treat a monochrome camera, we omit in Equation 7.2 the subscript k used in Equation 6.4.

are very large. Note in particular the logical inversion for patches 5 and 6 (blue flower and bluish green). We also tried to model the system as an ideal camera (flat spectral response) with an equienergetic illuminant, but the predicted values with this model were even further from the observed ones, see Figure 7.5.

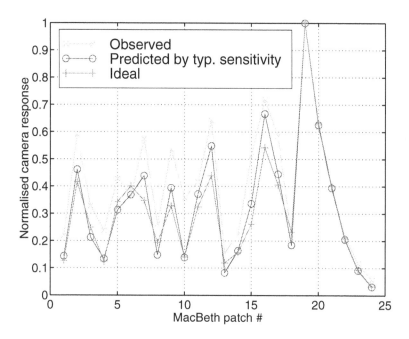

Figure 7.5: *Comparison of real experimental camera response on the Mac-Beth patches to the predicted response using two linear models, one with the camera spectral sensitivity as given by the manufacturer and illuminant A, and another with an ideal flat spectral sensitivity and the equienergetic illuminant. These preliminary models show poor performance.*

This preliminary experiment made us aware of a very important factor, namely the wavelength range used in our model compared to the real wavelength range in which the spectral sensitivity is non-zero. This is illustrated in Figure 7.6. The camera response to a given target patch is proportional to the area under the curve defined by the multiplication of the spectral sensitivity of the camera, the spectral radiance of the illuminant, and the spectral reflectance of the surface. If a reduced wavelength range is used in the model, *e.g.* from 400 to 700 nm, the error (red areas) becomes large.

These errors became extremely important when trying to estimate the spectral sensitivity as described in the next section. To avoid these problems, we decided to extend the wavelength range of 400 to 760 nm in our calculations, and to add an infrared (IR) cut-off filter in the optical path. It has a cut-off wavelength of about 720nm.

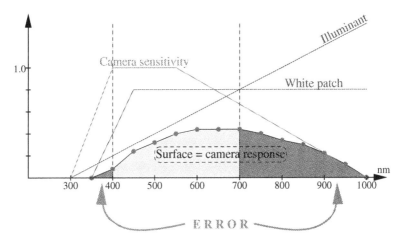

Figure 7.6: *Illustration of error induced when using a wavelength range of 400 to 700 nm. The difference between the observed and estimated camera response is very large. This error may be reduced in two ways: by extending the model's wavelength range, and by reducing the range in which the camera is sensitive, typically by adding an infrared cut-off filter.*

7.4.2   Estimation results

We applied the two methods proposed in Section 6.2.2 to estimate the spectral sensitivity ω (camera+illuminant, *cf.* Eq. 7.2), namely the Pseudoinverse (PI) and the Principal Eigenvector (PE) methods, on the experimental data, that is, on the normalized mean pixel values $I_n(i,j)$ of the patches of the Macbeth chart. Note that we would have preferred to use a color chart of carefully selected Munsell patches, as proposed in Section 6.2.2.5, but we did not dispose of the necessary patches to do this. The observed camera responses of the P patches of the chart with spectral reflectances $\mathbf{R}$ can be expressed as $\mathbf{c}_P = \mathbf{R}^t \boldsymbol{\omega} + \boldsymbol{\epsilon}$, cf. Equation 6.5 on page 127. After estimating the spectral sensitivity $\tilde{\boldsymbol{\omega}}$ we may then simulate the camera response as $\tilde{\mathbf{c}}_P = \mathbf{R}^t \tilde{\boldsymbol{\omega}}$.

As expected, the PI method performed very poorly. We show in Figure 7.7 the results of the PE method with the number of Principal Eigenvectors, r, varying from 1 to 6. For each value of the parameter r, the estimated sensitivity is shown. In Figure 7.8 we show a comparison of the experimentally observed camera responses and those predicted using the model with the different estimated sensitivities. We report the RMS ratio $\|\mathbf{c}_P - \tilde{\mathbf{c}}_P\|/\|\mathbf{c}_P\|$ between the observed and predicted values, and observe that the RMS ratio decreases with increasing r. However, it is clear that the estimate becomes poor when $r \geq 6$. As will be justified later in this chapter, we choose the estimate obtained with $r = 5$ for the further analysis.

7.5 Experimental multispectral image acquisition

As a basis for further analysis we have performed an acquisition of a 17-channel multispectral image of the Macbeth chart, varying the peak wavelength of the tunable filter in 20 nm steps from 400 nm to 720 nm. Nine channels of this multispectral image are shown in Figure 7.9. By selecting subsets of these images, we can simulate multispectral image acquisition with different numbers of channels. A tungsten halogen lamp driven by 4.0A/10.4V, a CCD binning of (H2, V2) giving a resolution of 640×512 pixels, and an aperture of f/2.8 was used. The integration times were chosen individually for each of the channels so as to yield a maximum digital signal without causing signal clipping, see Table 7.1. Then we corrected these images for the illuminant and the dark current as described in Section 7.3. The image-dependent normalization factors k_I (cf. Eq. 7.1) for each of the channels are also given in Table 7.1.

Worth noting is the particularly low integration time used for the 400 nm filter. We had expected this integration time to be higher than for the 420 nm filter since both the spectral distribution of the illuminant and the spectral transmittance of the LCTF filter decrease for decreasing peak wavelength. However, this behavior can be explained from the fact that

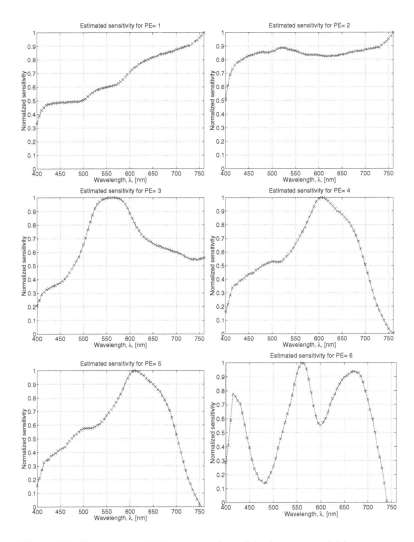

Figure 7.7: *Spectral sensitivity estimation of the image acquisition system consisting of a PCO SVGA SensiCam CCD camera and a tungsten halogen illuminant, using the PE method with $r = 1 \ldots 6$ Principal Eigenvectors and the Macbeth color chart. The estimated sensitivity obtained with $r = 5$ has been chosen for the further experimentation.*

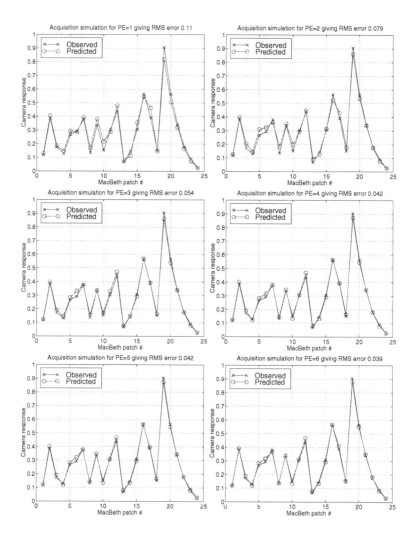

Figure 7.8: *Comparison of estimated and observed camera responses using the different estimations of spectral sensitivity given in Figure 7.7.*

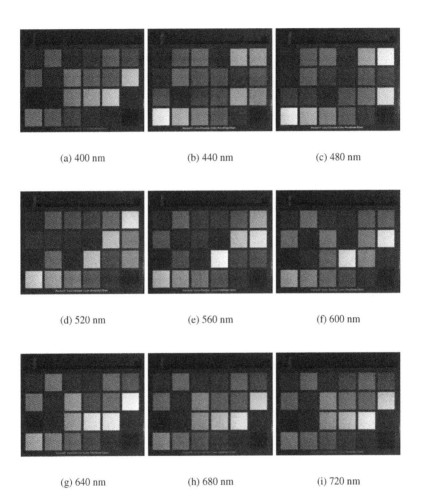

(a) 400 nm   (b) 440 nm   (c) 480 nm

(d) 520 nm   (e) 560 nm   (f) 600 nm

(g) 640 nm   (h) 680 nm   (i) 720 nm

Figure 7.9: *Nine channels of a multispectral image of the Macbeth color checker using the PCO SensiCam CCD camera and the LCT filter with varying peak wavelengths.*

the 400 nm filter has a considerable side-lobe around 700 nm (see Section 7.2.2). This un-wanted side-lobe also explains why the acquired images for 400 and 700 nm are very similar (see Figure 7.9)

Table 7.1: *Integration times t_k and normalization factors k_{I_k} for the 17 channels of our experimental multispectral image acquisition of the Macbeth chart.*

Peak wavelength [nm]			400	420	440	460
Integration time t_k [ms]			1000	6000	2000	1300
Normalization factor k_{I_k}			0.7452	0.3122	0.3022	0.3212
480	500	520	540	560	580	600
700	400	250	200	170	140	120
0.3254	0.3354	0.3659	0.3941	0.4220	0.4689	0.5005
620	640	660	680	700	720	
80	80	70	70	90	130	
0.4736	0.4614	0.5447	0.6341	0.6756	0.7517	

7.5.1 Model evaluation

We will now establish a linear model for the multispectral image acquisition based on the theory of Section 6.3, taking into account the integration times and the normalization factors. The vector $\mathbf{c}_K = [c_1 c_2 \ldots c_K]^t$ representing the response to all K filters (after normalization) may be described as

$$\mathbf{c}_K = \mathbf{A}\Theta^t\mathbf{r}, \tag{7.3}$$

where Θ is the known matrix of filter transmittances multiplied by the estimated spectral sensitivity, *i.e.* the matrix element of Θ is

$$\theta_{kn} = \phi_k(\lambda_n)\omega(\lambda_n), \quad 1 \le k \le K, \quad 1 \le n \le N.$$

The matrix $\mathbf{A}$ consists of the weights a_{kk} (see Table 7.1) on the diagonal, and zeros elsewhere:

$$a_{kk} = \alpha k_{I_k} t_k \tag{7.4}$$

The common normalization factor α is introduced in the model to be able to work with relative measurements of the spectral sensitivity, spectral power distribution of the illuminant, etc. It is determined by minimizing the RMS camera response estimation error.

Using this model, we can estimate the camera response to the Macbeth patches for each of the 17 filters, and we compare the estimates to the observed camera responses. We perform this simulation using six different estimations of the spectral sensitivity, obtained by a PE(r) estimation with the number of PE's r varying from 1 to 6, *cf.* Figure 7.7. The results of this simulation are shown in Figure 7.10. By examining the overall RMS ratio

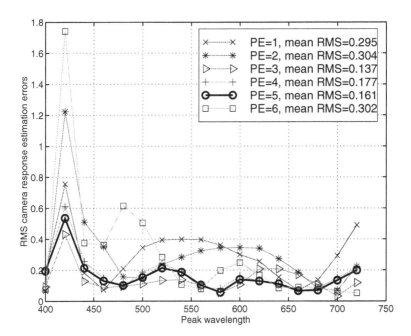

Figure 7.10: *RMS camera response estimation errors over the 24 Macbeth patches using the 17 different filters. We have also included in the figure the mean RMS of these over the 17 camera channels, and we note that $r = 3, 4, 5$ might be good solutions. Note the relationship between high errors for given wavelengths, and obvious errors in acquisition system sensitivity estimations (Figure 7.7), especially for $r = 6$.*

corresponding to the estimation errors for different choices of r, as defined in Section 7.4.2, we see that $r = 3, 4, 5$ gives reasonably small estimation errors, with a minimum of 0.137 for $r = 3$. However, we know that an IR cut-off filter is present in the optical path, and we can thus exclude

$r = 3$ since it has an important sensitivity in the red end of the spectrum, *cf.* Figure 7.7. We confirm therefore the choice of $r = 5$ as the optimal parameter for the spectral sensitivity estimation, *cf.* Section 7.4.2.

Having chosen PE(5) as the spectral sensitivity we perform a further comparison of the observed and estimated camera responses for the 24 Macbeth patches using the 17 different filters, as shown in Figure 7.11. We see that the differences between the observed and estimated camera responses are relatively large, especially for the 420 nm filter. These results are not satisfactory. We see from the figure that the errors are not randomly distributed. For a given filter there is often a tendency to either over- or under-estimation. We propose thus to modify the normalization matrix $\mathbf{A}$ in the model of Equation 7.3 to allow for independent normalization of each channel, that is, we redefine a_{kk} as compared to Equation 7.4, as

$$a_{kk} = \alpha_k k_{I_k} t_k, \tag{7.5}$$

and we choose the normalization factors α_k such that the camera response estimation errors *for each channel* are minimized. By using this modified model, we reduce the mean RMS camera estimation error by more than a factor 2, from 0.161 to 0.077, see Figure 7.12. The use of separate normalization factors α_k for each channel is mainly justified from the fact that we have limited confidence in the spectral sensitivity estimation.

7.6 Recovering colorimetric and spectrophotometric image data

We now examine how colorimetric and spectrophotometric information can be determined from the camera responses, *cf.* Section 6.3.

7.6.1 Model-based spectral reconstruction

Given the camera responses $\mathbf{c}_K = \mathbf{A}\Theta^t\mathbf{r}$ (*cf.* Equation 7.3) for a given surface, our goal is to estimate the spectral reflectance of the surface by using a reconstruction matrix $\mathbf{Q}$ as given by Equation 6.16, $\tilde{\mathbf{r}} = \mathbf{Q}\mathbf{c}_K$.

The simple pseudo-inverse solution of Section 6.3.1, $\mathbf{Q}_0 = (\mathbf{A}\Theta^t)^-$, being abandoned, we tried to use the method described in Section 6.3.2, exploiting *a priori* statistical spectral information of the imaged objects,

$$\mathbf{Q}_1 = \mathbf{R}\mathbf{R}^t\Theta\mathbf{A}(\mathbf{A}\Theta^t\mathbf{R}\mathbf{R}^t\Theta\mathbf{A})^{-1}.$$

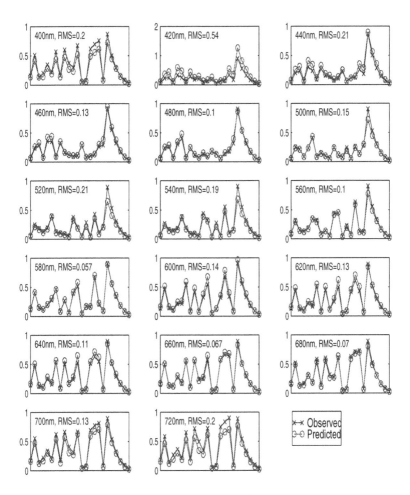

Figure 7.11: *Observed and predicted camera responses for the 24 Macbeth patches using the 17 different filters and the spectral sensitivity estimation PE(5). The overall mean RMS camera estimation ratio is 0.161.*

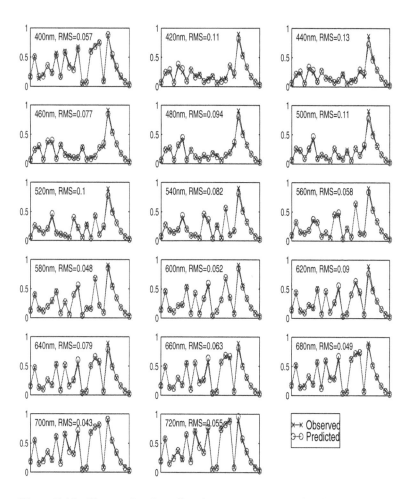

Figure 7.12: *Observed and predicted camera responses for the 24 Macbeth patches using the 17 different filters and the spectral sensitivity estimation PE(5), and seperate α_k for each channel. The overall mean RMS camera estimation ratio is 0.077.*

It became rapidly clear from the tests that the unmodified $\mathbf{Q}_1$ method did not give satisfactory results. This is due to the relatively important deficiencies of our model in estimating the camera output, as seen for example in Figure 7.11. The problem of estimating the spectral reflectance given the camera outputs and spectral sensitivities is very much similar to the problem of estimating the spectral sensitivity given the camera outputs and spectral reflectances. Both are inverse problems, and noise is present in both systems. We propose thus to apply a Principal Eigenvector (PE) approach for the estimation of spectral reflectances similar to the one presented in Section 6.2.2.4 for the spectral sensitivity estimation.

Four cases were defined, using 3, 6, 9, and 17 channels, defined by the following filter sets:

- K=3: $\{460, 560, 660\}$ nm,

- K=6: $\{400, 460, 520, 580, 640, 700\}$ nm,

- K=9: $\{400, 440, \ldots, 680, 720\}$ nm, and

- K=17: $\{400, 420, \ldots, 700, 720\}$ nm.

For each case, we evaluate the mean RMS spectral reconstruction error[5] while varying the parameter r. We evaluate this model-based estimation using two variants of the acquisition model, with a global normalization factor α or with individual ones α_k for each channel, as described in the previous section.

The results for the first variant of the model are shown in Figure 7.13(a). We see that due to the high level of noise only $r = 3$ Principal Eigenvectors can be used in the reconstruction process, no improvement of the results being achieved by using more than three filters (see Figure 7.13(b)). When too many Eigenvectors are taken into account in the system inversion, the noise severely deteriorates the estimation results.

For the modified model, the results are significantly better, as expected. The mean RMS spectral reconstruction errors for the four filter sets varying the parameter r is showed in Figure 7.14(a). For the 9 and 17 filter sets, a minimal mean RMS spectral reconstruction error of 0.24 is attained with the PE(5) estimation method. We achieve thus a much better spectral reconstuction (see Figure 7.14(b)).

---

[5]Note that the RMS spectral reconstruction error corresponds to the Euclidean distance in *spectral reflectance space, cf.* Section 6.6. It is *not* normalized, as it is the case for the

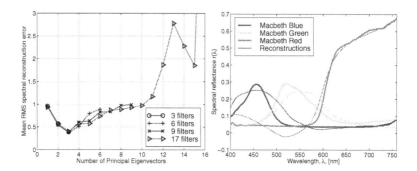

(a) Mean RMS spectral reconstruction errors for the four filter sets varying the parameter r

(b) Example of spectral reconstruction using the set of three filters and PE(3) reconstruction

Figure 7.13: *Spectral reconstruction of the Macbeth patches from the SensiCam camera responses with different filter sets using a Principal Eigenvector approach PE(r), with the original model.*

To gain more insight in how the PE(r) spectral reconstruction method succeeds in estimating reflectance spectra, we show in Figure 7.15 the spectral reflectance estimations of three of the Macbeth patches, for the set of 9 filters, varying r from 1 to 9. For PE(1) and PE(2) the dimensionality of the solution is too low, for PE(3) the reconstructions start to resemble the original spectra, for $r = 4$ to 7 they are quite good, but for 8 and 9 the estimations are slightly worse.

In Table 7.2 we resume the estimation results for different filter sets, and different values for r. We report the mean and maximal RMS difference between the original and reconstructed spectra. The differences are also expressed colorimetrically, in CIEXYZ and CIELAB color spaces (illuminant A) by applying standard formulae. An additional filter set marked 3', having peak wavelengths of 440, 560, and 600 nm, was chosen in order to be closer to the XYZ color matching functions (Figure 2.9 on page 25). Compared to the original 3-filter set the spectral errors are larger, while the

---

RMS *ratio* introduced earlier in this chapter to compare the observed and predicted *camera response values.*

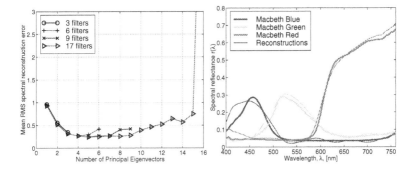

(a) Mean RMS spectral reconstruction errors for the four filter sets varying the parameter r using the modified model

(b) Example of spectral reconstruction using the set of nine filters and PE(5) reconstruction

Figure 7.14: *Spectral reconstruction of the Macbeth patches from the SensiCam camera responses with different filter sets using a Principal Eigenvector approach PE(r), with the modified model. The modified model with α_k normalization factors allows for reasonably good spectral reconstruction quality.*

colorimetric errors are smaller, as would be expected. We see that generally, using more filters gives smaller reconstruction errors. This result was in accordance with the simulations; see Table 6.8 on page 166. However, this trend is true only up to a certain number of filters; there is no significant improvement by using 17 instead of 9 filters. Results better than a mean ΔE_{ab} error of 3 are not obtained. The reason for this relatively poor performance when using many channels is mainly the fact that our spectral model of the image acquisition system does not predict the camera output values as precisely as we had hoped. The prediction of the spectral reflectances by model inversion then becomes somewhat hazardous. We will therefore in the next section evaluate an alternative way of using the camera output values.

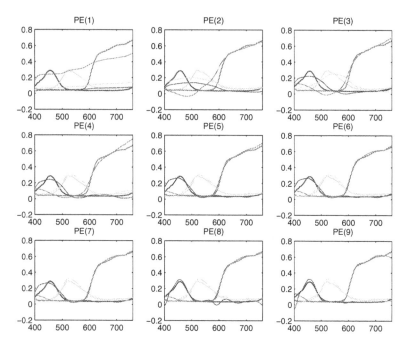

Figure 7.15: *Spectral reconstruction of three of the Macbeth patches from the Sensicam camera responses using the nine-filter set, and the PE(r) reconstruction method, with r varying from 1 to 9.*

7.6.2 Direct colorimetric regression

The idea here is to consider the acquisition system parameters as a completely unknown system, a *black box*, and simply try to recover directly the XYZ values from the camera output by regression. We have conducted two experiments, selecting sets of filters heuristically (Section 7.6.2.1) and optimally (Section 7.6.2.2).

7.6.2.1 Heuristical filter selection

In a first experiment we selected heuristically different subsets of the 17 channels to evaluate the method for different numbers of filters. Two sets of three and four filters were chosen with regards to the CIEXYZ color matching functions (Figure 2.9), while sets of 6, 7, 9, 15 and 17 filters

Table 7.2: *Mean and maximal errors expressed in spectral reflectance space, XYZ space and CIELAB space, for different filter sets and different spectral reconstruction methods. The filter set marked 3' was chosen in order to be closer to the XYZ color matching functions.*

# of filters	Recon. method	Spectral RMS		ΔXYZ		ΔE	
		Mean	Max	Mean	Max	Mean	Max
3	PE(3)	0.346	0.956	0.0264	0.0872	7.96	23.6
3'	PE(3)	0.507	1.466	0.0250	0.114	6.86	15.2
6	PE(3)	0.314	0.777	0.0268	0.0900	9.81	31.3
6	PE(5)	0.295	1.112	0.0221	0.111	4.42	13.1
9	PE(5)	0.244	0.782	0.0211	0.123	3.33	9.13
9	PE(9)	0.421	1.515	0.0211	0.117	3.44	8.75
17	PE(5)	0.243	0.729	0.0215	0.124	3.30	8.22
17	PE(9)	0.281	1.416	0.0213	0.119	3.08	8.56

were chosen to have equidistant peak wavelengths. The CIEXYZ tristimulus values were estimated by *linear* regression, and we report the mean Euclidean distance in XYZ space, $\overline{\Delta XYZ}$, as well as the mean and maximal ΔE_{ab} reconstruction errors taken under illuminant A in Table 7.3. The reconstructions using three and seven filters are illustrated in Figure 7.16. These results are quite as expected, *e.g.* when comparing our result using 6 filters ($\overline{\Delta E} = 4.4$), with Abrardo *et al.* (1996), who attains a mean ΔE error of 2.9 by linear regression from 6 camera channels to XYZ space using a subset of 20 patches of an AGFA IT8.7/3 color chart. It is however worth noting that the maximal error is greater using four filters than when using three. This may seem surprising. However, such effects are well-known when optimizing a RMS error. A minimal RMS error does not necessarily imply a minimal maximal error.

For the set of three filters, we also applied the non-linear method described in Section 3.2.3.5, in which 3rd order 3D polynomial regression is applied on the cubic root of the camera output (see the last line of Table 7.3). This method with *three* filters outperforms the linear regression with *seventeen* filters! It is however worth noting that the Macbeth chart is not well suited for this method, since the 20 coefficients of the polynomial is optimized using only 24 patches.

If we compare the results obtained by colorimetric linear regression (Table 7.3) with those obtained by model-based spectral reconstruction

(Table 7.2) for the same filter sets and the preferred (PE(5)) reconstruction methods, we can draw some interesting conclusions. For the sets of three (marked 3') and six filters, the results are nearly equivalent. For nine filters the colorimetric methods is slightly better. For the set of seventeen filters the colorimetric regression gives much smaller residual errors.

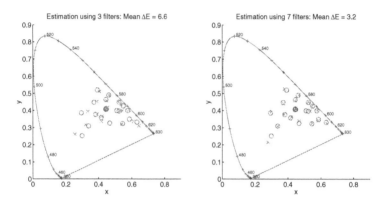

Figure 7.16: *Examples of colorimetric reconstructions using three and seven channel images and linear regression from camera responses to XYZ. The spectrally measured* (x, y) *values of the patches are marked with circles* $(\circ)$, *and those estimated from the acquired camera values with crosses* $(\times)$.

7.6.2.2   Filter selection by exhaustive search

We performed an evaluation of all possible combinations of three and four out of the seventeen filters. This required 680 comparisons for 3 filters and 2380 for 4. We then selected the combinations that minimized the mean errors in XYZ and CIELAB spaces as well as the maximal error in CIELAB space, and compared these results to our first heuristic selections. The results are reported in Table 7.4. We see that, especially for the 4-filter set, significantly better results are obtained by an exhaustive search, and the set of filters with peak wavelengths of 440, 460, 540, and 600 nm would seem a very good choice.

Table 7.3: *Resulting colorimetric reconstruction errors (using CIE illuminant A) using regression methods and different numbers of filters.*

#	Wavelengths	Method	$\overline{\Delta XYZ}$	$\overline{\Delta E}$	ΔE_{max}
3'	440 560 600	Linear to XYZ	0.0244	6.60	16.4
4	440 520 560 600	Linear to XYZ	0.0234	6.38	24.8
6	400 460 ... 640 700	Linear to XYZ	0.0182	4.40	13.8
7	440 480 520 ... 640 680	Linear to XYZ	0.0130	3.20	8.01
9	400 440 480 ... 680 720	Linear to XYZ	0.0124	2.56	8.41
15	420 440 460 ... 680 700	Linear to XYZ	0.00588	1.26	2.42
17	400 420 440 ... 700 720	Linear to XYZ	0.00345	0.95	2.91
3'	440 560 600	3rd ord. to LAB	—	0.63	1.91

Table 7.4: *Resulting reconstruction errors (illuminant A) using combinatorial search to select the best set of filters among the 17 compared to the heuristic selections of Table 7.3.*

#	Wavelengths	Select. meth.	Recon. meth.	$\overline{\Delta XYZ}$	$\overline{\Delta E}$	ΔE_{max}
3	440 560 600	Heuristic	Lin. XYZ	0.0244	6.60	16.4
3	460 540 600	Min. $\overline{\Delta XYZ}$	Lin. XYZ	0.0218	4.64	12.51
3	460 540 600	Min. $\overline{\Delta E}$	Lin. XYZ	idem	idem	idem
3	460 540 600	Min. ΔE_{max}	Lin. XYZ	idem	idem	idem
3	440 560 600	Heuristic	3rd ord. LAB		0.63	1.91
3	440 560 600	Min. $\overline{\Delta E}$	3rd ord. LAB		idem	idem
3	460 520 600	Min. ΔE_{max}	3rd ord. LAB		0.98	1.88
4	440 520 560 600	Heuristic	Lin. XYZ	0.0234	6.38	24.8
4	440 460 560 620	Min. $\overline{\Delta XYZ}$	Lin. XYZ	0.0165	5.94	26.0
4	460 500 540 620	Min. $\overline{\Delta E}$	Lin. XYZ	0.0254	3.63	9.18
4	440 460 540 600	Min. ΔE_{max}	Lin. XYZ	0.0175	3.67	8.65

7.7   Conclusion

The multispectral image acquisition experiment presented here has been enriching in several ways. First of all, it has reminded us that simulations and reality are two very different things. We have seen that the noise involved in the image acquisition process was much larger than the quantization noise used in the simulations (Sec. 6.2.2.2).

This has led us *i)* to propose a modified image acquisition model with separate normalization factors for each channel; *ii)* to propose a new method using a Principal Eigenvector approach for the estimation of a spectral reflectance given the camera responses through several filters; and *iii)* to consider using simpler regression techniques to obtain colorimetric information from the camera responses. Note however that by doing the latter, we only obtain information about the color of a surface under a given illuminant, not about its spectral reflectance as with the model-based spectral reconstruction method.

Several different ways to convert the output values of the camera into device-independent color or spectral reflectance information might be investigated and tested, for example to estimate the first K' principal components of a reflectance spectrum from the K camera output as proposed by Burns (1997), or to estimate the reflectance values for given wavelengths directly by interpolation between the camera responses using narrow-band filters of varying peak wavelengths, *e.g.* by spline interpolation or Modified Discrete Sine Transformation (Keusen and Praefcke, 1995, Keusen, 1996, König and Praefcke, 1998a).

As seen in Table 7.3 the residual ΔE_{ab} color differences are very much smaller when using non-linear regression to CIELAB space than using linear regression to XYZ space. This confirms the results obtained for a flatbed scanner in Chapter 3. A possible extension to improve further the results would be to use non-linear regression methods also with $K > 3$ filters. One would need to make sure that the number of patches of the color chart remains larger than the number of parameters of the model, though.

We have seen that the choice of filters is important. For example with four filters, the maximum ΔE_{ab} color difference was reduced from 24.8 to 8.65 by optimizing the filter selection. It would be interesting to pursue these tests to evaluate the different filter selection methods proposed in Section 6.5.

We conclude that the multispectral image acquisition system we have assembled presents several strong interests. The computer-controlled CCD camera and LCTF tunable filters are easy to use, and the colorimetric and spectrophotometric quality is quite good. Unfortunately, time has not allowed us to use this system as much as we would have preferred. One interesting application that we could have realized in practice is the simulation of a fine arts painting as seen under different illuminants, as described in Section 6.7.

Chapter 8

Conclusions and perspectives

Conclusions

In this research, we have worked on several problems, all aimed towards a common goal: To achieve high image quality throughout a digital imaging chain. We have been especially concerned with the quality of color acquisition and reproduction.

We have developed novel algorithms for the colorimetric characterization of scanners and printers, providing efficient and colorimetrically accurate means of conversion between a device-independent color space such as the CIELAB space, and the device-dependent color spaces of a scanner and a printer. The algorithms we have developed have been implemented and transferred to industry and are currently being used in commercial color management software. Some of the proposed methods are protected by a patent.

The scanner characterization process (Chapter 3) has been shown to introduce a significant improvement of the colorimetric quality of the image acquisition performed by the scanner. A mean ΔE_{ab} color difference of 1 between the colors of the color chart and the colors estimated from the scanner RGB values is achieved. This difference is hardly perceptible. This method and several other methods for correcting for image artifacts have been applied to achieve very high quality in image acquisition of fine art paintings (Chapter 4).

The proposed printer characterization method (Chapter 5) presents several strong points of interest. First, it performs efficiently the transformation from CIELAB (or any other 3D color space) to CMY directly without additional numerical optimization techniques. Secondly it is able to easily incorporate different gamut mapping techniques, both continuous and clipping methods. Thirdly it is versatile, not being limited to a specific printing technology.

To go a step further in image quality and fidelity, we have developed algorithms for multispectral image capture using a CCD camera with carefully selected optical filters (Chapter 6). A new method for the selection of a reduced number of spectral samples to be used for the estimation of the spectral sensitivity functions of an electronic camera has been developed. This optimized selection method allowed to get virtually the same performance using only 20 patches, compared to using the complete set of 1269 Munsell chips. Several methods have been developed and tested for the choice of filters and for the estimation of the spectral reflectance, taking into account the statistical spectral properties of the objects that are to be imaged, as well as the spectral characteristics of the camera and the spectral radiance of the illuminant that is used for the acquisition. We have shown that multispectral image data present several advantages compared to conventional color images, in particular for simulating a scene as seen under a given illuminant.

Finally, an experimental multispectral camera was assembled using a professional monochrome CCD camera and an optical tunable filter (Chapter 7), validating in practice the theoretical models and simulations of the previous chapter. To be able to recover colorimetric and spectrophotometric information about the imaged surface from the camera output signals, two main approaches were proposed. One consists in applying an extended version of the colorimetric scanner characterization method described above to convert from the camera outputs to a device-independent color space such as CIEXYZ or CIELAB. Another method is based on the spectral model of the acquisition system. By inverting the model, we can estimate the spectral reflectance of each pixel of the imaged surface.

To conclude, we feel that the work described in this dissertation has been interesting in several aspects. It contains elements from several fields of science, such as signal and image processing, computer science, applied mathematics, physics, and color science, the latter being itself indeed an interdisciplinary field. But the different algorithms developed in this work have also been successfully utiliz—ed for several applications, such as fine-arts archiving, color facsimile, and color management systems, and some of them have been transfered to the industry.

Perspectives

Time did not allow us to develop all of the many ideas that were suggested in the discussions around this thesis and numerous important topics remain to be addressed. In the following, we try to indicate what seems to us to be interesting trails to follow starting from the results of this work.

- It would be interesting to apply the algorithm of printer characterization to several types of printing systems, to verify its versatility and robustness. The development of algorithms for automatic detection and elimination of erroneous vertices would be helpful.

- An extension of the printer characterization algorithm from three to four or more inks would also be of great interest. Multi-ink printing is a field of research and development that is in rapid growth (see *e.g.* MacDonald *et al.*, 1994, Herron, 1996, Van De Capelle and Meireson, 1997, Mahy and DeBaer, 1997, Berns *et al.*, 1998, Tzeng and Berns, 1998). Note that if there exists a unique invertible transformation from CMY to a multi-ink representation, the current method may be used. If this is not the case, a more complex geometric structure should be developed.

- It could be interesting to evaluate the colorimetric quality of a color facsimile system, that is, by comparing the colors of an original document with a facsimile of this document obtained by successive scanning and printing. Several iterations of this process could be made. The results of such an experiment depend not only on the quality of the colorimetric characterization algorithms, but also on the choice of gamut mapping method.

- The proposed methods for colorimetric characterizations of scanners and printers require a sufficiently large number of color patches to give accurate results. This might be a practical limitation, especially for printer characterization. A very interesting research subject would be to investigate the use of a simpler method of *calibration* requiring only a small number of color patches, to perform minor adjustments of the geometrical structure determined by the proposed characterization procedure.

▓ The capability of the printer characterization method to incorporate different types of color gamut mapping methods could be exploited for the design and evaluation of such methods. Psychophysical tests with human observers should then be effectuated, to be able to quantify the resulting image quality after gamut mapping (Morovic, 1998).

▓ To further improve the degree of color consistency between different media, color appearance models could be used instead of the CIELAB space for data exchange. Several authors have proposed color appearance models (CAM), *e.g.* RLAB by Fairchild and Berns (1993) or the Hunt model (Chapter 31 of Hunt, 1995). Refer to Fairchild (1997) for a comprehensive presentation of existing CAMs. Recently, the CIE agreed on a standardized color appearance model, the CIECAM97s (CIE TC1-34, 1998). This model would be a natural choice for further work.

▓ Concerning the comparative statistical analysis of different sets of spectral reflectances, it would be very interesting to evaluate several proposed measures to compare the vector space spanned by the different sets of spectral reflectances. We would then seek to find correlations between the different measures and the quality of spectral reconstruction when an encoding adapted for one data set is applied to another.

▓ For the spectral characterization of image acquisition devices, the results could be further improved by adding constraints on the estimations such as smoothness and positivity, for example by using the Wiener estimation method (Pratt and Mancill, 1976), the technique of projection onto convex sets (POCS) (Sharma and Trussell, 1993; 1996c), or quadratic programming (Finlayson *et al.*, 1998).

Furthermore, special consideration could be made for representing illuminants having spiky spectral radiances, such as fluorescent lamps. Such illuminants are very common in flatbed scanners. One possibility is to represent such spectra by a combination of smooth basis vectors and ray spectra as done *e.g.* by Sharma and Trussell (1996c).

Further experiments could also be done to evaluate the possible gain obtained by taking into account *a priori* information about the spectral distribution of the illuminant, *i.e.* by measuring it spectrophotometrically.

It would be very interesting to compare the spectral sensitivity curves obtained by our indirect method to those obtained by direct measurements using a device that emits wavelength-tunable monochromatic light.

▨ To evaluate the quality of a match between an observed spectral reflectance and one estimated from camera responses, more sophisticated measures than the Euclidean spectral distance could be used, such as the Goodness-Fitting-Coefficients proposed by Romero *et al.* (1997) and García-Beltrán *et al.* (1998).

▨ The experimental multispectral camera consisting of a monochrome CCD camera and a LCTF tunable filter constitutes a very powerful tool for color and multispectral image acquisition. This should be exploited in the future. Further testing, research and software development is required in order to improve its quality and usability. Objectives may be to develop an imaging spectrophotometer or colorimeter of high accuracy, for example by simulating CIEXYZ color matching functions by a linear combination of the camera sensitivities of K channels. The limits of attainable accuracy may be studied (Burns, 1997).

Digital color imaging is a research field with great prospects. Many problems are still unsolved. We hope that the work described in this dissertation may serve the community, both through developed methods, and by giving ideas for further work concerning the technology, science, and art of color.

Bibliography

Abrardo, A., Cappellini, V., Cappellini, M., and Mecocci, A. (1996). Art-works colour calibration using the VASARI scanner. In *Proceedings of IS&T and SID's 4th Color Imaging Conference: Color Science, Systems and Applications*, pages 94–97, Scottsdale, Arizona.

Albert, A. (1972). *Regression and the Moore-Penrose pseudoinverse.* Academic Press.

Allain, C. (1989). Caractérisation d'une caméra Eikonix. Rapport de stage, École Nationale Supérieure des Télécommunications.

Anderson, M., Motta, R., Chandrasekar, S., and Stokes, M. (1996). Proposal for a standard default color space for the internet — sRGB. In *Proceedings of IS&T and SID's 4th Color Imaging Conference: Color Science, Systems and Applications*, pages 238–246, Scottsdale, Arizona. Updated version 1.10 can be found at http://www.color.org/sRGB.html.

ANSI IT8.7/2 (1993). American National Standard IT8.7/2-1993: Graphic technology - Color reflection target for input scanner calibration. NPES, Reston, Virginia.

ANSI IT8.7/3 (1993). American National Standard IT8.7/3-1993: Graphic technology - Input data for characterization of 4-color process printing. NPES, Reston, Virginia.

Arai, Y., Nakauchi, S., and Usui, S. (1996). Color correction method based on the spectral reflectance estimation using a neural network. In *Proceedings of IS&T and SID's 4th Color Imaging Conference: Color Science, Systems and Applications*, pages 5–9, Scottsdale, Arizona.

Balasubramanian, R. (1994). Color transformations for printer color correction. In *Proceedings of IS&T and SID's 2nd Color Imaging Conference: Color Science, Systems and Applications*, pages 62–65, Scottsdale, Arizona.

Bell, I. E. and Cowan, W. (1993). Characterizing printer gamuts using tetrahedral interpolation. In *Proceedings of IS&T and SID's Color Imaging Conference: Transforms and Transportability of Color*, pages 108–113, Scottsdale, Arizona.

Ben Jemaa, R. (1998). *Traitement de données 3D denses acquises sur des objets réels complexes*. Ph.D. thesis, École Nationale Supérieure des Télécommunications.

Beretta, G. (1996). Image processing for color facsimile. In *Proceedings of the 5th International Conference on High Technology: Imaging Science and Technology, Evolution and Promise*, pages 148–156, Chiba, Japan.

Bern, M. and Eppstein, D. (1992). Mesh generation and optimal triangulation. In Du and Wang (1992), pages 23–90.

Berns, R. S. (1993a). Colorimetric characterization of Sharp JX610 desktop scanner. Technical report, Munsell Color Science Laboratory.

Berns, R. S. (1993b). Spectral modeling of a dye diffusion thermal transfer printer. Technical report, Munsell Color Science Laboratory.

Berns, R. S. (1998). Challenges for color science in multimedia imaging. In *Conference Proceedings, CIM'98, Colour Imaging in Multimedia*, pages 123–133, Derby, UK.

Berns, R. S., Motta, R. J., and Gorzynski, M. E. (1993). CRT colorimetry. part I: Theory and practice. *Color Research and Application*, **18**(5), 299–314.

Berns, R. S., Imai, F. H., Burns, P. D., and Tzeng, D.-Y. (1998). Multi-spectral-based color reproduction research at the Munsell Color Science Laboratory. In *Electronic Imaging: Processing, Printing, and Publishing in Color*, volume 3409 of *SPIE Proceedings*, pages 14–25.

Billmeyer, F. W. and Saltzman, M. (1981). *Principles of Color Technology*. John Wiley, New York, 2 edition.

Borouchaki, H. (1993). *Graphe de connexion et triangulation de Delaunay*. Ph.D. thesis, Université de PARIS VII.

Borouchaki, H., Norguet, F., and Schmitt, F. (1994). Graphe de connexion et triangulation dans R^d. *C. R. Acad. Sci. Paris*, **318**(1), 283–288.

Bournay, E. (1991). *Calibration et acquisition multifiltres à l'aide de la caméra Eikonix.* Master's thesis, École Nationale Supérieure des Télécommunications.

Bowyer, A. (1981). Computing Dirichlet tessellations. *The Computer Journal,* **2**(24), 162–166.

Braun, K. M. and Fairchild, M. D. (1997). Testing five color-appearance models for changes in viewing conditions. *Color Research and Application,* **22**(3), 165–173.

Brettel, H., Chiron, A., Hardeberg, J. Y., and Schmitt, F. (1997). Versatile spectrophotometer for cross-media color management. In *Proceedings of the 8th Congress of the International Colour Association, AIC Color 97,* volume II, pages 678–681, Kyoto, Japan.

Burger, R. E. and Sherman, D. (1994). Producing colorimetric data from film scanners using a spectral characterization target. In *Device Independent Color Imaging,* volume 2170 of *SPIE Proceedings,* pages 42–52.

Burns, P. D. (1997). *Analysis of image noise in multispectral color acquisition.* Ph.D. thesis, Center for Imaging Science, Rochester Institute of Technology.

Burns, P. D. and Berns, R. S. (1996). Analysis multispectral image capture. In *Proceedings of IS&T and SID's 4th Color Imaging Conference: Color Science, Systems and Applications,* pages 19–22, Scottsdale, Arizona.

Camus-Abonneau, A. and Camus, C. (1989). Analyse spectrophotométrique et modèle numérique d'enregistrement de tableaux. Rapport de stage, École Polytechnique and École Nationale Supérieure des Télécommunications.

Chang, I. C. (1976). Tunable acusto-optic filters: An overview. In *Acousto-Optics: Device Development/Instrumentation/Application,* volume 90 of *SPIE Proceedings,* pages 12–22.

Chang, Y.-L., Liang, P., and Hackwood, S. (1989). Unified study of color sampling. *Applied Optics,* **28**(4), 809–813.

Chen, P. and Trussell, H. J. (1995). Color filter design for multiple illuminants and detectors. In *Proceedings of IS&T and SID's 3rd Color Imaging Conference: Color Science, Systems and Applications,* pages 67–70, Scottsdale, Arizona.

Chrien, T., Chovit, C., and Miller, P. (1993). Imaging spectrometry using liquid crystal tunable filters. From http://stargate.jpl.nasa.gov/lctf/.

CIE 15.2 (1986). *Colorimetry*, volume 15.2 of *CIE Publications*. Central Bureau of the CIE, Vienna, Austria, 2 edition.

CIE 17.4 (1989). *International Lighting Vocabulary*, volume 17.4 of *CIE Publications*. Central Bureau of the CIE, Vienna, Austria.

CIE 80 (1989). *Special Metamerism Index: Change in Observer*, volume 80 of *CIE Publications*. Central Bureau of the CIE, Vienna, Austria.

CIE TC1-34 (1998). The CIE 1997 interim colour appearance model (simple version), CIECAM97s. *CIE Division 1 Final Draft*, **TC1-34**. http://www.cis.rit.edu/people/faculty/fairchild/PDFs/CIECAM97s.pdf.

Clarke, F. J. J., McDonald, R., and Rigg, B. (1984). Modification to the jpc79 colour-difference formula. *J. Soc. Dyers & Colorists*, **100**, 128–132.

Cohen, J. (1964). Dependency of the spectral reflectance curves of the Munsell color chips. *Psychonomic Science*, **1**, 369–370.

Crettez, J.-P. (1998). Analyse colorimétrique des peintures : représentation des couleurs dans l'espace CIELAB 1976. *Infovisu, SID France*, pages 12–15.

Crettez, J.-P. and Hardeberg, J. Y. (1999). Analyse colorimétrique des peintures : étude comparative de 3 tableaux de J.-B. Corot. *TECHNE, Laboratoire de Recherche des Musées de France*, (9–10), 52–60.

Crettez, J.-P., Hardeberg, J. Y., Maître, H., and Schmitt, F. (1996). Numérisation de huit tableaux à haute résolution et haute qualité colorimétrique. *85 œuvres du Musée du Louvre - Analyse scientifique, CD-ROM AV 500041, Conception LRMF, Réunion des Musées de France*.

Cupitt, J. (1996). Methods for characterising digital cameras. In R. Luo, editor, *Colour Management for Information Systems*. University of Derby, Derby, GB.

Dannemiller, J. L. (1992). Spectral reflectance of natural objects: how many basis functions are necessary. *Journal of the Optical Society of America A*, **9**(4), 507–515.

Deconinck, G. (1990). *Multi-channel Images: Acquisition and Coding Problems*. M.Sc. thesis, Katholieke Universiteit Leuven and École Nationale Supérieure des Télécommunications.

Delaunay, B. (1934). Sur la sphère vide. *Izv. Akad. Nauk SSSR Otdelenie Matemat. Estestvennyka Nauk*, **7**, 793–800.

Dongarra, J. J., Bunch, J. R., Moler, C. B., and Stewart, G. W. (1979). *LINPACK User's Guide*. SIAM, Philadelphia.

Dress, W. H., Tenkolsky, S. A., Vetterling, W. T., and Flannery, B. P. (1992). *Numerical Recipies in C. The Art of Scientific Computing*. Cambridge University Press, 2 edition.

Du, D.-Z. and Wang, F. K. (1992). *Computing in Euclidian Geometry*. World Scientific Publishing Co., 2 edition.

Eem, J. K., Shin, H. D., and Park, S. O. (1994). Reconstruction of surface spectral reflectances using characteristic vectors of Munsell colors. In *Proceedings of IS&T and SID's 2nd Color Imaging Conference: Color Science, Systems and Applications*, pages 127–131, Scottsdale, Arizona.

Engeldrum, P. G. (1993). Color scanner colorimetric design requirements. In *Device-Independent Color Imaging and Imaging Systems Integration*, volume 1909 of *SPIE Proceedings*, pages 75–83.

Fairchild, M. D. (1997). *Color Appearance Models*. Addison-Wesley Publishing Company.

Fairchild, M. D. and Berns, R. S. (1993). Image color-appearance specification through extension of CIELAB. *Color Research and Application*, **18**(3), 178–190.

Farkas, D. L., Ballou, B. T., Fisher, G. W., Fishman, D., Garini, Y., Niu, W., and Wachman, E. S. (1996). Microscopic and mesoscopic spectral bio-imaging. In *Proc. Soc. Photo-Optical Instr. Eng*, volume 2678, pages 200–206.

Farrell, J. E. and Wandell, B. A. (1993). Scanner linearity. *Journal of Electronic Imaging*, **3**, 147–161.

Farrell, J. E., Sherman, D., and Wandell, B. A. (1994). How to turn your scanner into a colorimeter. In *Proceedings of IS&T Tenth International Congress on Advances in Non-Impact Printing Technologies*, pages 579–581.

Finlayson, G. D., Hubel, P. M., and Hordley, S. (1997). Color by correlation. In *Proceedings of IS&T and SID's 5th Color Imaging Conference: Color Science, Systems and Applications*, pages 6–11, Scottsdale, Arizona.

Finlayson, G. D., Hordley, S., and Hubel, P. M. (1998). Recovering device sensitivities with quadratic programming. In *Proceedings of IS&T and SID's 6th Color Imaging Conference: Color Science, Systems and Applications*, Scottsdale, Arizona.

Fortune, S. (1992). Voronoi diagrams and Delaunay triangulations. In Du and Wang (1992), pages 193–233.

Fumoto, T., Kanamori, K., Yamada, O., Motomura, H., Kotera, H., and Inoue, M. (1995). SLANT/PRISM convertible structured color processor MN5515. In *Proceedings of IS&T and SID's 3rd Color Imaging Conference: Color Science, Systems and Applications*, pages 101–105, Scottsdale, Arizona.

Funt, B. (1995). Linear models and computational color constancy. In *Proceedings of IS&T and SID's 3rd Color Imaging Conference: Color Science, Systems and Applications*, pages 26–29, Scottsdale, Arizona.

García-Beltrán, A., Nieves, J. L., Hernández-Andrés, J., and Romero, J. (1998). Linear bases for spectral reflectance functions of acrylic paints. *Color Research and Application*, **23**(1), 39–45.

Gentile, R. S., Walowitt, E., and Allebach, J. P. (1990). A comparison of techniques for color gamut mismatch compensation. *Journal of Imaging Technology*, **16**(5), 176–181.

Golub, G. H. and Reinsch, C. (1970). Singular value decomposition and least squares solutions. *Numer. Math.*, **14**, 403–420.

Golub, G. H. and Reinsch, C. (1971). Singular value decomposition and least squares solutions. In J. H. Wilkinson and C. Reinsch, editors, *Linear Algebra*, Handbook for Automatic Computation, chapter I/10, pages 134–151. Springer-Verlag.

Golub, G. H. and van Loan, C. F. (1983). *Matrix Computations*. The Johns Hopkins University Press.

Goulam-Ally, F. (1990a). Application informatique: reconstruction de spectres à partir de données caméra. Rapport de stage, École Nationale Supérieure des Télécommunications.

Goulam-Ally, F. (1990b). Calibration spectrophotométrique d'une camera multispectrale. Rapport de stage, École Nationale Supérieure des Télécommunications.

Grassmann, H. G. (1853). Zur theorie der farbenmischung. *Annalen der Physik und Chemie*, **89**, 69. English translation "Theory of Compound Colors" published in Philosophical Magazine, vol 4(7), pp. 254–264 (1854). Reprinted in MacAdam (1993), pp 10–13.

Haneishi, H., Hirao, T., Shimazu, A., and Miyake, Y. (1995). Colorimetric precision in scanner calibration using matrices. In *Proceedings of IS&T and SID's 3rd Color Imaging Conference: Color Science, Systems and Applications*, pages 106–108, Scottsdale, Arizona.

Haneishi, H., Hasegawa, T., Tsumura, N., and Miyake, Y. (1997). Design of color filters for recording art works. In *Proceedings of IS&T 50th Annual Conference*, pages 369–372.

Hardeberg, J. Y. (1995). *Transformations and Colour Consistency for the Colour Facsimile*. Diploma thesis, The Norwegian Institute of Technology (NTH), Trondheim, Norway.

Hardeberg, J. Y. and Crettez, J.-P. (1998). Computer aided colorimetric analysis of fine art paintings. In *Oslo International Colour Conference, Colour between Art and Science*, Oslo, Norway.

Hardeberg, J. Y. and Schmitt, F. (1996). Colorimetric characterization of a printer for the color facsimile. Technical report, ENST / SEPT.

Hardeberg, J. Y. and Schmitt, F. (1997). Color printer characterization using a computational geometry approach. In *Proceedings of IS&T and SID's 5th Color Imaging Conference: Color Science, Systems and Applications*, pages 96–99, Scottsdale, Arizona. Also in R. Buckley, ed., *Recent Progress in Color Management and Communications*, IS&T, pages 88-91, 1998.

Hardeberg, J. Y. and Schmitt, F. (1998). Colour management: Why and how. In *Oslo International Colour Conference, Colour between Art and Science*, pages 135–136, Oslo, Norway.

Hardeberg, J. Y., Schmitt, F., Tastl, I., Brettel, H., and Crettez, J.-P. (1996). Color management for color facsimile. In *Proceedings of IS&T and SID's 4th Color Imaging Conference: Color Science, Systems and Applications*, pages 108–113, Scottsdale, Arizona. Also in R. Buckley, ed., *Recent Progress in Color Management and Communications*, IS&T, pages 243-247, 1998.

Hardeberg, J. Y., Schmitt, F., Brettel, H., Crettez, J.-P., and Maître, H. (1998a). Multispectral imaging in multimedia. In *Conference Proceedings, CIM'98, Colour Imaging in Multimedia*, pages 75–86, Derby, UK.

Hardeberg, J. Y., Brettel, H., and Schmitt, F. (1998b). Spectral characterisation of electronic cameras. In *Electronic Imaging: Processing, Printing, and Publishing in Color*, volume 3409 of *SPIE Proceedings*, pages 100–109.

Hardeberg, J. Y., Schmitt, F., Brettel, H., Crettez, J.-P., and Maître, H. (1999). Multispectral image acquisition and simulation of illuminant changes. In L. W. MacDonald and R. Luo, editors, *Colour Imaging: Vision and Technology*, pages 145–164. John Wiley & Sons, Ltd.

Harding, R. W. (1997). Hyperspectral imaging: How much is hype? *Photonics Spectra*, **31**(6), 82–94.

Hauta-Kasari, M., Wang, W., Toyooka, S., Parkkinen, J., and Lenz, R. (1998). Unsupervised filtering of Munsell spectra. In *ACCV'98, 3rd Asian Conference on Computer Vision*, Hong Kong.

Hedley, M. and Yan, H. (1992). Segmentation of color images using spatial and color space information. *Journal of Electronic Imaging*, **1**(4), 374–380.

Hering, E. (1905). Grundzüge der Lehre vom Lichtsinn. *Handb. d. ges. Augenheilk.*, **XII**.

Hermeline, F. (1982). Triangulation automatique d'un polyèdre en dimension n. *RAIRO Modèl. Math. Anal. Numér*, **16**(3), 211–242.

Herron, S. (1996). An exploration of the Pantone Hexachrome six-color system reproduced by stochastic screens. In *Proceedings of IS&T and SID's 4th Color Imaging Conference: Color Science, Systems and Applications*, pages 114–120, Scottsdale, Arizona.

Hill, B. (1998). Multispectral color technology: A way towards high definition color image scanning and encoding. In *Electronic Imaging: Processing, Printing, and Publishing in Color*, volume 3409 of *SPIE Proceedings*, pages 2–13.

Horn, B. K. P. (1984). Exact reproduction of color images. *Comput. Vis., Graphics, Image Processing*, **26**, 135–167.

Hoshino, T. and Berns, R. S. (1993). Color gamut mapping techniques for color hard copy images. In *Device-Independent Color Imaging and Imaging Systems Integration*, volume 1909 of *SPIE Proceedings*, pages 152–165.

Hård, A., Sivik, L., and Tonnquist, G. (1996). NCS, natural color system — from concept to research and applications. part I and II. *Color Research and Application*, **21**(3), 180–220.

Hubel, P. M., Sherman, D., and Farrell, J. E. (1994). A comparison of methods of sensor spectral sensitivity estimation. In *Proceedings of IS&T and SID's 2nd Color Imaging Conference: Color Science, Systems and Applications*, pages 45–48, Scottsdale, Arizona.

Hung, P.-C. (1991). Colorimetric calibration for scanners and media. In *Camera and Input Scanner Systems*, volume 1448 of *SPIE Proceedings*, pages 164–174.

Hung, P.-C. (1993). Colorimetric calibration in electronic imaging devices using a look-up-table method and interpolations. *Journal of Electronic Imaging*, **2**(1), 53–61.

Hunt, R. W. G. (1991). *Measuring Colour*. Ellis Horwood, Hertfordshire, UK, 2 edition.

Hunt, R. W. G. (1995). *The Reproduction of Colour*. Fountain Press, Kingston-upon-Thames, UK, 5 edition.

Hurlbert, A. C., Wen, W., and Poggio, T. (1994). Learning colour constancy. *Journal of Photographic Science*, **42**, 89–90.

ICC.1:1998.9 (1998). File format for color profiles. The International Color Consortium. See http://www.color.org/.

IEC 61966-2.1 (1999). Colour measurement and management in multimedia systems and equipment, Part 2-1: Colour management - Default RGB colour space - sRGB. *IEC Publication*, **61966-2.1**.

IEC 61966-8 (1999). Colour measurement and management in multimedia systems and equipment, Part 8: Colour scanners. *IEC Committee Draft*, **61966-8**.

Iino, K. and Berns, R. S. (1998a). Building color management modules using linear optimization I. Desktop color system. *Journal of Imaging Science and Technology*, **42**(1).

Iino, K. and Berns, R. S. (1998b). Building color management modules using linear optimization II. Prepress system for offset printing. *Journal of Imaging Science and Technology*, **42**(2).

Imai, F. H., Tsumura, N., Haneishi, H., and Miyake, Y. (1996a). Principal component analysis of skin color and its application to colorimetric color reproduction on CRT display and hardcopy. *Journal of Imaging Science and Technology*, **40**(5), 422–430.

Imai, F. H., Tsumura, N., Haneishi, H., and Miyake, Y. (1996b). Spectral reflectance of skin color and its applications to color appearance modeling. In *Proceedings of IS&T and SID's 4th Color Imaging Conference: Color Science, Systems and Applications*, pages 121–124, Scottsdale, Arizona.

ISO 12639 (1997). Graphic technology — Prepress digital data exchange — Colour targets for input scanner calibration. *ISO TC 130 Standard*, **12639**.

ITU-T T.42 (1994). Continuous-tone colour representation method for facsimile. *ITU-T Recommendation*, **T.42**.

Ives, H. E. (1915). The transformation of color-mixture equations from one system to another. *J. Franklin Inst.*, **16**, 673–701.

Jaaskelainen, T., Parkkinen, J., and Toyooka, S. (1990). Vector-subspace model for color representation. *Journal of the Optical Society of America A*, **7**(4), 725–730. See http://www.it.lut.fi/research/color/database/database.html.

Johnson, T. (1992). Techniques for reproducing images in different media: Advantages and disadvantages of current methods. In *Proceedings of the Technical Association of the Graphic Arts (TAGA)*, pages 739–755.

Johnson, T. (1996). Methods for characterizing colour scanners and digital cameras. *Displays*, **14**(4), 183–191.

Jolliffe, I. T. (1986). *Principal Component Analysis*. Springer-Verlag, New York.

Judd, D. B. (1933). Sensibility to color-temperature change as a function of temperature. *Journal of the Optical Society of America*, **23**, 7–14. Reprinted in MacAdam (1993), pp 208–215.

Judd, D. B. and Wyszecki, G. (1975). *Color in Business, Science and Industry*. John Wiley, New York, 3 edition.

Kaiser, P. K. and Boynton, R. M. (1996). *Human Color Vision*. Optical Society of America, Washington, DC.

Kanamori, K., Kawakami, H., and Kotera, H. (1990). A novel color transformation algorithm and its applications. In *Image Processing Algorithms and Techniques*, volume 1244 of *SPIE Proceedings*, pages 272–281.

Kang, H. R. (1992). Color scanner calibration. *Journal of Imaging Science and Technology*, **36**(2), 162–170.

Kang, H. R. (1997). *Color Technology for Electronic Imaging Devices*. SPIE Optical Engineering Press.

Kasson, J. M., Nin, S. I., Plouffe, W., and Hafner, J. L. (1995). Performing color space conversions with threedimensional linear interpolation. *Journal of Electronic Imaging*, **4**(3), 226–250.

Katoh, N. and Ito, M. (1996). Gamut mapping for computer generated images II. In *Proceedings of IS&T and SID's 4th Color Imaging Conference: Color Science, Systems and Applications*, pages 126–129, Scottsdale, Arizona.

Keusen, T. (1994). Optimierte Auswertung Multispektraler Abtastsignale. In *DfwG-Tagung*, Illmenau.

Keusen, T. (1996). Multispectral color system with an encoding format compatible with the conventional tristimulus model. *Journal of Imaging Science and Technology*, **40**(6), 510–515.

Keusen, T. and Praefcke, W. (1995). Multispectral color system with an encoding format compatible to conventional tristimulus model. In *Proceedings of IS&T and SID's 3rd Color Imaging Conference: Color Science, Systems and Applications*, pages 112–114, Scottsdale, Arizona.

Khotanzad, A. and Zink, E. (1996). Segmentation of color maps using eigenvector line-fitting techniques. In *Proceedings of IS&T and SID's 4th Color Imaging Conference: Color Science, Systems and Applications*, pages 129–134, Scottsdale, Arizona.

König, F. and Praefcke, W. (1998a). A multispectral scanner. In *Conference Proceedings, CIM'98, Colour Imaging in Multimedia*, pages 63–73, Derby, UK.

König, F. and Praefcke, W. (1998b). The practice of multispectral image acquisition. In *Electronic Imaging: Processing, Printing, and Publishing in Color*, pages 34–41.

Kollarits, R. V. and Gibbon, D. C. (1992). Improving the color fidelity of cameras for advanced television systems. In *High-Resolution Sensors and Hybrid Systems*, volume 1656 of *SPIE Proceedings*, pages 19–29.

Kopp, G. and Derks, M. (1997). Stay tuned: Photonic filters color your world. *Photonics Spectra*, **31**(3), 125–130.

Kowaliski, P. (1990). *Vision et mesure de la couleur*. Masson et Cie. 2nd ed. updated by F. Viénot and R. Sève.

Krinov, E. L. (1947). Spectral'naye otrazhatel'naya sposobnost'prirodnykh obrazovanii. In *Izd. Akad. Nauk USSR (Proc. Acad. Sci. USSR)*. Translation by G. Belkov, "Spectral reflectance properties of natural formations," Technical Translation TT-439 (National Research Council of Canada, Ottawa, Canada, 1953).

Kroh, O. (1921). Über Farbkonstans und Farbentransformation. *Zeits. f. Sinnesphysiol.*, **52**, 181–216.

Kubelka, P. and Munk, F. (1931). Ein Beitrag zur Optik der Farbanstriche. *Z. tech. Phys.*, **12**, 593–601.

LeGrand, Y. (1957). *Light, Color, and Vision*. John Wiley, New York.

Lenz, R. and Lenz, U. (1989). ProgRes 3000: A digital color camera with a 2D array CCD sensor and programmable resolution up to 2994 x 2320 picture elements. In *MedTech'89: Medical Imaging*, volume 1357 of *SPIE Proceedings*, pages 204–209.

Lenz, R., Österberg, M., Hiltunen, J., Jaaskelainen, T., and Parkkinen, J. (1995). Unsupervised filtering of color spectra. In *Proceedings of the 9th Scandinavian Conference on Image Analysis*, pages 803–810, Uppsala, Sweden.

Lenz, R., Österberg, M., Hiltunen, J., Jaaskelainen, T., and Parkkinen, J. (1996a). Unsupervised filtering of color spectra. *Journal of the Optical Society of America A*, **13**(7), 1315–1324.

Lenz, U., Habil, D., and Lenz, R. (1996b). Digital camera color calibration and characterisation. In *Proceedings of IS&T and SID's 4th Color Imaging Conference: Color Science, Systems and Applications*, pages 23–24, Scottsdale, Arizona.

Lo, M.-C., Luo, M. R., and Rhodes, P. A. (1996). Evaluating colour models' performance between monitor and print images. *Color Research and Application*, **21**(4), 277–291.

Luo, M. R. and Morovič, J. (1996). Two unsolved issues in colour management — colour appearance and gamut mapping. In *Proceedings of the 5th International Conference on High Technology: Imaging Science and Technology, Evolution and Promise*, pages 136–147, Chiba, Japan.

Luo, M. R. and Rigg, B. (1987a). BFD(l:c) color-difference formula. part I — development of the formula. *J. Soc. Dyers Colorists*, **103**, 86–94.

Luo, M. R. and Rigg, B. (1987b). BFD(l:c) color-difference formula. part II — performance of the formula. *J. Soc. Dyers Colorists*, **103**, 126–132.

Luther, R. (1927). Aus dem Gebiet der Farbreizmetrik. *Z. Technol. Phys.*, **8**, 540–558.

Lyot, B. (1933). Un monochromateur à grand champ utilisant les interférences en lumière polarisée. *Compt. Rend.*, **197**, 1593.

MacAdam, D., editor (1993). *Selected Papers on Colorimetry — Fundamentals*, volume 77 of *Milestone Series*. SPIE Optical Engineering Press, Bellingham, Washington.

MacDonald, L. and Morovič, J. (1995). Assessing the effects of gamut compression in the reproduction of fine art paintings. In *Proceedings of IS&T and SID's 3rd Color Imaging Conference: Color Science, Systems and Applications*, pages 194–200, Scottsdale, Arizona.

MacDonald, L. W. (1993a). Device independent colour reproduction. In *Eurodisplay*, pages B–3/1–B–3/36.

MacDonald, L. W. (1993b). Gamut mapping in perceptual colour space. In *Proceedings of IS&T and SID's Color Imaging Conference: Transforms and Transportability of Color*, pages 193–196, Scottsdale, Arizona.

MacDonald, L. W. and Lenz, R. (1994). An ultra-high resolution digital camera. *Journal of Photographic Science*, **42**(2), 49–51.

MacDonald, L. W., Deane, J. M., and Rughani, D. N. (1994). Extending the colour gamut of printed images. *Journal of Photographic Science*, **42**, 97–99.

Mahy, M. and DeBaer, D. (1997). HIFI color printing within a color management system. In *Proceedings of IS&T and SID's 5th Color Imaging Conference: Color Science, Systems and Applications*, pages 277–283, Scottsdale, Arizona.

Mahy, M. and Delabastita, P. (1996). Inversion of the Neugebauer equations. *Color Research and Application*, **21**(6), 404–411.

Mahy, M., van Eycken, L., and Oosterlinck, A. (1994a). Evaluation of uniform color spaces developed after the adoption of CIELAB and CIELUV. *Color Research and Application*, **19**(2), 105–121.

Mahy, M., Wambacq, P., Eycken, L. V., and Oosterlinck, A. (1994b). Optimal filters for the reconstruction and discrimination of reflectance curves. In *Proceedings of IS&T and SID's 2nd Color Imaging Conference: Color Science, Systems and Applications*, pages 140–143, Scottsdale, Arizona.

Maillot, J. (1991). *Trois Approches du Plaquage de Texture sur un Objet Tridimensionnel*. Ph.D. thesis, Univ. de Paris VI.

Maître, H., Schmitt, F., Crettez, J.-P., Wu, Y., and Hardeberg, J. Y. (1996). Spectrophotometric image analysis of fine art paintings. In *Proceedings of IS&T and SID's 4th Color Imaging Conference: Color Science, Systems and Applications*, pages 50–53, Scottsdale, Arizona.

Maître, H., Schmitt, F., and Crettez, J.-P. (1997). High quality imaging in museum: from theory to practice. In *Very High Resolution and Quality Imaging II*, volume 3025 of *SPIE Proceedings*, pages 30–39, San Jose, California.

Maloney, L. T. (1986). Evaluation of linear models of surface spectral reflectance with small numbers of parameters. *Journal of the Optical Society of America A*, **3**(10), 1673–1683.

Maloney, L. T. and Wandell, B. A. (1986). Color constancy: a method for recovering surface spectral reflectance. *Journal of the Optical Society of America A*, **3**(1), 29–33.

Manabe, Y. and Inokuchi, S. (1997). Recognition of material types and analysis of interreflection using spectral image. In *Proceedings of the 8th Congress of the International Colour Association, AIC Color 97*, volume II, pages 507–510, Kyoto, Japan.

Manabe, Y., Sato, K., and Inokuchi, S. (1994). An object recognition through continuous spectral images. In *Proceedings of the 12th International Conference on Pattern Recognition*, volume I, PA4.33, pages 858–860.

Marcu, G. and Abe, S. (1995). Three-dimensional histogram visualization in different color spaces and applications. *Journal of Electronic Imaging*, **4**(4), 330–346.

Martinez, K., Cupitt, J., and Saunders, D. (1993). High resolution colorimetric imaging of paintings. In *Cameras, Scanners and Image Acquisition Systems*, volume 1901 of *SPIE Proceedings*, pages 25–36.

Martínez-Verdú, F., Pujol, J., and Capilla, P. (1998). Designing a tristimulus colorimeter from a conventional machine vision system. In *Conference Proceedings, CIM'98, Colour Imaging in Multimedia*, pages 319–333, Derby, UK.

Matlab Language Reference Manual (1996). *MATLAB, The Language of Technical Computing, Language Reference Manual*. The MathWorks, Inc. Version 5, See http:/www.mathworks.com.

Maxwell, J. C. (1857). The diagram of colors. *Transactions of the Royal Society of Edinburgh*, **21**, 275–298. Excerpts reprinted in MacAdam (1993), pp 17 19.

McCamy, C. S., Marcus, H., and Davidson, J. G. (1976). A color rendition chart. *Journal of Applied Photographic Engineering*, **2**, 95–99.

McDonald, R. and Smith, K. J. (1995). CIE94 — a new colour-difference formula. *J. Soc. Dyers Col.*, **111**, 376–379.

McLaren, K. (1986). *The Colour Science of Dyes and Pigments*. Hilger, Bristol, UK, 2 edition.

Miyake, Y. and Yokoyama, Y. (1998). Obtaining and reproduction of accurate color images based on human perception. In *Color Imaging: Device-Independent Color, Color Hardcopy, and Graphic Arts III*, volume 3300 of *SPIE Proceedings*, pages 190–197.

Müller, M. (1989). Calibration photométrique et colorimétrique d'une caméra haute résolution. Rapport de stage, École Nationale Supérieure des Télécommunications.

Montag, E. D. and Fairchild, M. D. (1997). Psychophysical evaluation of gamut mapping techniques using simple rendered images and artificial gamut boundaries. *IEEE Transactions on Image Processing*, **6**(7), 977–989.

Morovic, J. (1998). *To Develop a Universal Gamut Mapping Algorithm*. Ph.D. thesis, Colour & Imaging Institute, University of Derby.

Morovic, J. and Luo, M. R. (1997). Cross-media psychophysical evaluation of gamut mapping algorithms. In *Proceedings of the 8th Congress of the International Colour Association, AIC Color 97*, volume II, pages 594–597, Kyoto, Japan.

Morovic, J. and Luo, M. R. (1998). A universal algorithm for colour gamut mapping. In *Conference Proceedings, CIM'98, Colour Imaging in Multimedia*, pages 169–177, Derby, UK.

Motomura, H., Fumoto, T., Yamada, O., Kanamori, K., and Kotera, H. (1994). CIELAB to CMYK color conversion by prism and slant prism interpolation method. In *Proceedings of IS&T and SID's 2nd Color Imaging Conference: Color Science, Systems and Applications*, pages 156–159, Scottsdale, Arizona.

Müller, M. and Burmeister, A. (1993). Registration of transportation damages using a high resolution CCD camera. In *Recording Systems: High-Resolution Cameras and Recording Devices and Laser Scanning and Recording Systems*, volume 1987 of *SPIE Proceedings*, pages 111–117.

Munsell (1976). *Munsell Book of Color - Matte Finish Collection*. Munsell Color, Baltimore, Md.

Murch, G. (1993). Color management on the desktop. In *Proceedings of IS&T and SID's Color Imaging Conference: Transforms and Transportability of Color*, pages 95–99, Scottsdale, Arizona.

Nagel, B. (1993). Multispectral image reconstruction for visual art archiving. Rapport de stage, École Nationale Supérieure des Télécommunications.

Nassau, K. (1983). *The Physics and Chemistry of Color. The Fifteen Causes of Color.* John Wiley & sons.

Neugebauer, H. E. J. (1937). Die theoretischen Grundlagen des Mehrfarbendruckes. *Zeitschrift für wissenschaftliche Photographie Photophysik und Photochemie,* **36**(4).

Neugebauer, H. E. J. (1956). Quality factor for filters whose spectral transmittances are different from color mixture curves, and its application to color photography. *Journal of the Optical Society of America,* **46**, 821–824.

Newton, I. (1671). New theory about light and colors. *Philosophical Transactions of the Royal Society of London,* **80**, 3075–3087. Excerpts reprinted in MacAdam (1993), pp 3–4.

Newton, I. (1730). *Opticks.* London, UK, 4 edition. Reprinted by Dover, New York, 1979.

Nicodemus, F. E., Richmond, J. C., Hsia, J. J., Ginsberg, I. W., and Limperis, T. (1977). Geometrical considerations and nomenclature for reflectance. NBS Monogram 160, National Bureau of Standards.

Nimeroff, I. (1972). Colorimetry. In *Precision Measurement and Calibration, Selected NBS Papers on Colorimetry,* volume 9, pages 157–203.

NPL QM 117 (1995). The colorimetry of visual displays. NPL Report QM 117, The National Physical Laboratory.

Ohta, N. (1993). Development of color targets for input scanner calibration. In A. Nemcsics and J. Schanda, editors, *Proceedings of the 7th Congress of the International Colour Association,* volume B, pages 125–129, Budapest, Hungary.

Ohta, N. and Wyszecki, G. (1975). Theoretical chromaticity-mismatch limits of metamers viewed under different illuminants. *Journal of the Optical Society of America,* **65**, 327–333.

Pariser, E. G. (1991). An investigation of color gamut reduction techniques. In *IS&T Symposium on Electronic Prepress Technology - Color Printing,* pages 105–107.

Park, S. O., Kim, H. S., Park, J. M., and Eem, J. K. (1995). Development of spectral sensitivity measurement system of image sensor devices. In *Proceedings of IS&T and SID's 3rd Color Imaging Conference: Color Science, Systems and Applications,* pages 115–118, Scottsdale, Arizona.

Parkkinen, J. and Jaaskelainen, T. (1987). Color representation using statistical pattern recognition. *Applied Optics*, **26**(19), 4240–4245.

Parkkinen, J., Hallikainen, J., and Jaaskelainen, T. (1989). Characteristic spectra of Munsell colors. *Journal of the Optical Society of America A*, **6**, 318–322. See http://www.it.lut.fi/research/color/database/database.html.

Peercy, M. S. (1993). Linear color representations for full speed spectral rendering. *Proc. 20th annual conference on Computer graphics, SIGGRAPH'93*, pages 191–198.

Photo CD (1991). Kodak Photo CD System. A Planning Guide for Developers, Eastman Kodak Company, Part No. DCI200R. See also http://www.kodak.com/go/photocd/.

Pointer, M. R. and Attridge, G. G. (1998). The number of discernible colours. *Color Research and Application*, **23**, 52–54. See also page 337 of the same volume.

Poynton, C. (1996). *A Technical Introduction to Digital Video*. John Wiley & Sons,, 1 edition. See also the author's *Frequently-Asked Questions about Gamma* http://www.inforamp.net/~poynton/GammaFAQ.html and *Frequently-Asked Questions about Color* http://www.inforamp.net/~poynton/ColorFAQ.html.

Praefcke, W. (1996). Transform coding of reflectance spectra using smooth basis vectors. *Journal of Imaging Science and Technology*, **40**(6), 543–548.

Praefcke, W. and Keusen, T. (1995). Optimized basis functions for coding reflectance spectra minimizing the visual color difference. In *Proceedings of IS&T and SID's 3rd Color Imaging Conference: Color Science, Systems and Applications*, pages 37–40, Scottsdale, Arizona.

Pratt, W. K. (1978). *Digital Image Processing*. John Wiley & sons.

Pratt, W. K. and Mancill, C. E. (1976). Spectral estimation techniques for the spectral calibration of a color image scanner. *Applied Optics*, **15**(1), 73–75.

Preparata, F. P. and Shamos, M. I. (1985). *Computational Geometry: an Introduction*. Springer-Verlag.

Rajala, S. A. and Kakodkar, A. P. (1993). Interpolation of color data. In *Proceedings of IS&T and SID's Color Imaging Conference: Transforms and Transportability of Color*, pages 180–183, Scottsdale, Arizona.

Rao, A. R. (1998). Color calibration of a colorimetric scanner using non-linear least squares. In *Proc. IS&T's 1998 PICS Conference*, Portland, OR.

Romero, J., García-Beltrán, A., and Hernández-Andrés, J. (1997). Linear bases for representation of natural and artificial illuminants. *Journal of the Optical Society of America A*, **14**(5), 1007–1014.

Rosselet, A., Graff, W., Wild, U. P., Keller, C. U., and Gschwind, R. (1995). Persistent spectral hole burning used for spectrally high-resolved imaging of the sun. In *Imaging Spectrometry*, volume 2480 of *SPIE Proceedings*, pages 205–212.

Sato, T., Nakano, Y., Iga, T., and Usui, S. (1997). Color reproduction using low dimensional spectral reflectance. In *Proceedings of the 8th Congress of the International Colour Association, AIC Color 97*, volume II, pages 553–556, Kyoto, Japan.

Saunders, D. and Cupitt, J. (1993). Image processing at the National Gallery: The VASARI project. *National Gallery Technical Bulletin*, (14).

Savin, T. (1998). Calibrage d'un filtre spectral accordable à cristaux liquides. Rapport de stage, Université Denis Diderot (Paris 7) et École Nationale Supérieure des Télécommunications.

Sayanagi, K. (1986). Black printer UCR and UCA. In *Proceedings of the Technical Association of the Graphic Arts (TAGA)*, pages 402–429.

Schläpfer, K., Steiger, W., and Grönberg, J. (1998). Features of color management systems. UGRA Report 113/1, Association for the Promotion of Research in the Graphic Arts Industry.

Schmitt, F. (1976). A method for the treatment of metamerism in colorimetry. *Journal of the Optical Society of America*, **66**, 601–608.

Schmitt, F. (1996). High quality digital color images. In *Proceedings of the 5th International Conference on High Technology: Imaging Science and Technology, Evolution and Promise*, pages 55–62, Chiba, Japan.

Schmitt, F. and Hardeberg, J. Y. (1997). Transformation colorimétrique pour dispositif de restitution de couleurs. Brevet français No. 97-05331, France Telecom SA.

Schmitt, F. and Hardeberg, J. Y. (1998). Colorimetric transformation for colour-restoring device. European patent EP874229A1. France Telecom SA.

Schmitt, F., Maître, H., and Wu, Y. (1990). First progress report: tasks 2.4 (Development / procurement of basic software routines) and 3.3 (Spectrophotometric characterization of paintings) — Vasari project. Technical Report 2649, CEE ESPRIT II.

Schmitt, F., Wu, Y., Crettez, J.-P., and Boulay, G. (1995). Color calibration for color facsimile. In *SID International Symposium*, Orlando.

Schmitt, F., Crettez, J.-P., Brettel, H., Hardeberg, J. Y., and Tastl, I. (1996). Input and output device characterization in the field of color facsimile. In *Proceedings of the CIE Expert Symposium: Colour Standards for Image Technology*, volume CIE x010-1996, pages 141–143, Vienna, Austria.

SensiCam Specification (1997). Sensicam: Specifications and typical values. PCO Computer Optics GmbH. See also http://www.pco.de/English/Products/ Specs/dsht_sc.htm.

Sharma, G. and Trussell, H. J. (1993). Characterization of scanner sensitivity. In *Proceedings of IS&T and SID's Color Imaging Conference: Transforms and Transportability of Color*, pages 103–107, Scottsdale, Arizona.

Sharma, G. and Trussell, H. J. (1996a). Measures of goodness for color scanners. In *Proceedings of IS&T and SID's 4th Color Imaging Conference: Color Science, Systems and Applications*, pages 28–32, Scottsdale, Arizona.

Sharma, G. and Trussell, H. J. (1996b). Optimal filter design for multi-illuminant color correction. In *Proc. IS&T/OSA's Optics and Imaging in the Information Age*, pages 83–86, Rochester, NY.

Sharma, G. and Trussell, H. J. (1996c). Set theoretic estimation in color scanner characterization. *Journal of Electronic Imaging*, **5**, 479–489.

Sharma, G. and Trussell, H. J. (1997a). Digital color imaging. *IEEE Transactions on Image Processing*, **6**(7), 901–932.

Sharma, G. and Trussell, H. J. (1997b). Figures of merit for color scanners. *IEEE Transactions on Image Processing*, **6**(7), 990–1001.

Sharma, G., Wang, S., Sidavanahalli, D., and Knox, K. (1998). The impact of UCR on scanner calibration. In *Proc. IS&T's 1998 PICS Conference*, pages 121–124, Portland, OR.

Sherman, D. and Farrell, J. E. (1994). When to use linear models for color calibration. In *Proceedings of IS&T and SID's 2nd Color Imaging Conference: Color Science, Systems and Applications*, pages 33–36, Scottsdale, Arizona.

Shiobara, T., Zhou, S., Haneishi, H., Tsumura, N., and Miyake, Y. (1995). Improved color reproduction of electronic endoscopes. In *Proceedings of IS&T and SID's 3rd Color Imaging Conference: Color Science, Systems and Applications*, pages 186–190, Scottsdale, Arizona.

Shiobara, T., Zhou, S., Haneishi, H., Tsumura, N., and Miyake, Y. (1996). Improved color reproduction of electronic endoscopes. *Journal of Imaging Science and Technology*, **40**(6), 494–501.

Sivik, L. (1997). NCS — reflecting the color sense as a perceptual system. In *Proceedings of the 8th Congress of the International Colour Association, AIC Color 97*, volume I, pages 50–57, Kyoto, Japan.

Souami, F. (1993). *Traitement de la réflectance des objets dans un système d'acquisition 3D couleur*. Ph.D. thesis, ENST 93E033.

Souami, F. and Schmitt, F. (1995). Estimation de la réflectance d'une surface connaissant sa géométrie et sa couleur. *Traitement du Signal*, **12**(2), 145–158.

Spaulding, K. E., Ellson, R. N., and Sullivan, J. R. (1995). UltraColor: A new gamut mapping strategy. In E. Walowit, editor, *Proceedings of Device Independent Color Imaging II*, volume 2414 of *SPIE Proceedings*, pages 61–68.

Stockman, A., MacLeod, D. I. A., and Johnson, N. E. (1993). Spectral sensitivities of the human cones. *Journal of the Optical Society of America A*, **10**, 2491–2521.

Stokes, M. (1997). Industry adoption of color management systems. In *Proceedings of the 8th Congress of the International Colour Association, AIC Color 97*, volume I, pages 126–131, Kyoto, Japan.

Stone, M. C. and Wallace, W. E. (1991). Gamut mapping computer generated imagery. In *Graphics Interface*, pages 32–39.

Stone, M. C., Cowan, W. B., and Beatty, J. C. (1988). Color gamut mapping and the printing of digital color images. *ACM Transactions on Graphics*, **7**(4), 249–292.

Sugiura, H., Kuno, T., and Ikeda, H. (1998). Methods of measurement for colour reproduction of digital cameras. In *Digital Solid State Cameras: Designs and Applications*, volume 3302 of *SPIE Proceedings*, pages 113–122.

Sève, R. (1996). *Physique de la Couleur, de l'apparence colorée à la technique colorimétrique*. Masson, 1 edition.

Swain, P. H. and Davis, S. M., editors (1978). *Remote Sensing: The Quantitative Approach*. McGraw-Hill, New York.

Tajima, J. (1996). New quality measures for a set of color sensors — weighted quality factor, spectral characteristic restorability index and color reproducibility index —. In *Proceedings of IS&T and SID's 4th Color Imaging Conference: Color Science, Systems and Applications*, pages 25–28, Scottsdale, Arizona.

Tominaga, S. (1996). Multichannel vision system for estimating surface and illumination functions. *Journal of the Optical Society of America A*, **13**(11), 2163–2173.

Tominaga, S. (1997). Computational approach and a multi-channel vision system. In *Proceedings of the 8th Congress of the International Colour Association, AIC Color 97*, volume A, pages 71–76, Kyoto, Japan.

Trussell, H. J. (1991). Applications of set theoretic methods to color systems. *Color Research and Application*, **16**(1), 31–41.

Trussell, H. J. and Kulkarni, M. S. (1996). Sampling and processing of color signals. *IEEE Transactions on Image Processing*, **5**(4), 677–681.

Tsumura, N., Imai, F. H., Saito, T., Haneishi, H., and Miyake, Y. (1997). Color gamut mapping based on Mahalanobis distance for color reproduction of electronic endoscope image under different illuminant. In *Proceedings of IS&T and SID's 5th Color Imaging Conference: Color Science, Systems and Applications*, pages 158–162, Scottsdale, Arizona.

Tzeng, D.-Y. and Berns, R. S. (1998). Spectral-based ink selection for multiple-ink printing I. Colorant estimation of original object. In *Proceedings of IS&T and SID's 6th Color Imaging Conference: Color Science, Systems and Applications*, Scottsdale, Arizona.

Van De Capelle, J. P. and Meireson, B. (1997). A new method for characterizing output devices and its fit into ICC and HIFI color workflows. In *Proceedings of IS&T and SID's 5th Color Imaging Conference: Color Science, Systems and Applications*, pages 66–69, Scottsdale, Arizona.

von Kries, J. (1902). Chromatic adaptation. *Festschrift der Albrecht-Ludwigs-Universität*, pages 145–158. Excerpts reprinted in MacAdam (1993), pp 47–52.

Vora, P. L. and Trussell, H. J. (1993). Measure of goodness of a set of colour scanning filters. *Journal of the Optical Society of America A*, **10**(7), 1499–1508.

Vora, P. L. and Trussell, H. J. (1997). Mathematical methods for the design of color scanning filters. *IEEE Transactions on Image Processing*, **6**(2), 312–320.

Vora, P. L., Trussell, H. J., and Iwan, L. (1993). A mathematical method for designing a set of colour scanning filters. In *Color Hard Copy and Graphic Arts II*, volume 1912 of *SPIE Proceedings*, pages 322–332.

Vrhel, M. J. and Trussell, H. J. (1992). Color correction using principal components. *Color Research and Application*, **17**(5), 328–338.

Vrhel, M. J. and Trussell, H. J. (1994). Filter considerations in color correction. *IEEE Trans. Image Proc.*, **3**(2), 147–161.

Vrhel, M. J., Gershon, R., and Iwan, L. S. (1994). Measurement and analysis of object reflectance spectra. *Color Research and Application*, **19**(1), 4–9. See ftp://ftp.eos.ncsu.edu/pub/spectra.

Vrhel, M. J., Trussell, H. J., and Bosch, J. (1995). Design and realization of optimal color filters for multi-illuminant color correction. *Journal of Electronic Imaging*, **4**(1), 6–14.

Wandell, B. A. (1987). The synthesis and analysis of color images. *IEEE Transactions on Pattern Analysis and Machine Intelligence*, **9**(1), 2–13.

Wandell, B. A. (1995). *Foundations of Vision*. Sinauer, Sunderland, MA.

Wandell, B. A. and Farrell, J. E. (1993). Water into wine: Converting scanner RGB into tristimulus XYZ. In *Device-Independent Color Imaging and Imaging Systems Integration*, volume 1909 of *SPIE Proceedings*, pages 92–101.

Wang, W., Hauta-Kasari, M., and Toyooka, S. (1997). Optimal filters design for measuring colors using unsupervised neural network. In *Proceedings of the 8th Congress of the International Colour Association, AIC Color 97*, volume I, pages 419–422, Kyoto, Japan.

Watson, D. F. (1981). Computing the n-dimensional Delaunay tessellation with application to Voronoi polytopes. *Comput. J*, **24**, 167–172.

Wolski, M., Allebach, J. P., and Bouman, C. A. (1994). Gamut mapping. Squeezing the most out of your color system. In *Proceedings of IS&T and SID's 2nd Color Imaging Conference: Color Science, Systems and Applications*, pages 89–92, Scottsdale, Arizona.

Wu, Y., Crettez, J.-P., Schmitt, F., and Maître, H. (1991). Correction des distortions colorimétriques d'une caméra RVB. In *VISU-91*, pages 94–95, Perros Guirec, France.

Wyszecki, G. and Stiles, W. S. (1982). *Color Science: Concepts and Methods, Quantitative Data and Formulae*. John Wiley & Sons, New York, 2 edition.

Yokoyama, Y., Tsumura, N., Haneishi, H., Miyake, Y., Hayashi, J., and Saito, M. (1997). A new color management system based on human perception and its application to recording and reproduction of art paintings. In *Proceedings of*

IS&T and SID's 5th Color Imaging Conference: Color Science, Systems and Applications, pages 169–172, Scottsdale, Arizona.

Young, R. A. (1986). Principal-component analysis of macaque lateral geniculate nucleus chromatic data. *Journal of the Optical Society of America A*, **3**(10), 1735–1742.

Zaidi, Q. (1998). Identification of illuminant and object colors: heuristic-based algorithms. *Journal of the Optical Society of America A*, **15**, 1767–1776.

Citation index

Appendices

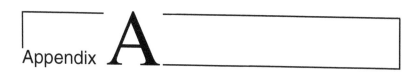

Appendix **A**

Mathematical background

A.1 Least mean square (LMS) error minimization

This section describes a solution for finding a polynomial approximation for a three-dimensional transformation, given a known set of corresponding points for that transformation. The method for finding this transformation is based on minimizing the mean square error (MSE) between all known theoretical function values y_i, and the calculated values $\varphi(x_i)$, $i = [0, \dots, m]$. $\varphi(\cdot)$ is called the interpolation function. The general expression for the MSE is given as:

$$\text{MSE} = \frac{1}{m+1} \sum_{i=0}^{m} (\varphi(x_i) - y_i)^2 \qquad (\text{A.1})$$

Before considering the 3D interpolation problem, the method of finding the mean square solution is introduced by a 1D interpolation problem.

A.1.1 1D Interpolation Functions of degree n

Our interpolation function may be a polynomial function of degree n as shown in Equation A.2. Supposing we have more values $y_i, i = 0, \dots, m$ than unknowns, that is $n + 1 < m$, it is generally impossible to find an interpolation function $\varphi(\cdot)$ that fits exactly into the given set of values y_i.

$$\varphi_n(x) = a_0 + a_1 x + \ldots + a_n x^n, \quad (a_k \in \mathbb{R}; k = 0, \ldots, n) \quad \text{(A.2)}$$

To minimize the MSE given in Equation A.1 on the page before, the coefficients $a_0, a_1, \ldots, a_n$ have to be chosen such that

$$F(a_0, a_1, \ldots, a_n) = \sum_{i=0}^{m} [a_0 + a_1 x_i + \ldots + a_n x^n - y_i]^2 \quad \text{(A.3)}$$

is minimized. In other words:

$$\frac{\partial F}{\partial a_k} = 2 \sum_{i=0}^{m} [a_0 + a_1 x_i + \ldots + a_n x^n - y_i] x_i^k = 0, \quad k = 0, \ldots, n \quad \text{(A.4)}$$

Equations A.4 generates a system of $n + 1$ equations to find the unknown coefficients $a_0, \ldots, a_n$ as follows:

$$a_0(m+1) + a_1 \sum_{i=0}^{m} x_i + \ldots + a_n \sum_{i=0}^{m} x_i^n = \sum_{i=0}^{m} y_i$$

$$a_0 \sum_{i=0}^{m} x_i + a_1 \sum_{i=0}^{m} x_i^2 + \ldots + a_n \sum_{i=0}^{m} x_i^{n+1} = \sum_{i=0}^{m} x_i\, y_i$$

$$\vdots$$

$$a_0 \sum_{i=0}^{m} x_i^n + a_1 \sum_{i=0}^{m} x_i^{n+1} + \ldots + a_n \sum_{i=0}^{m} x_i^{2n} = \sum_{i=0}^{m} x_i^n\, y_i \quad \text{(A.5)}$$

We introduce the following matrix notation to solve Equations A.5 :

$$\mathbf{y} = \begin{bmatrix} y_0 \\ y_1 \\ \vdots \\ y_m \end{bmatrix}, \quad \mathbf{a} = \begin{bmatrix} a_0 \\ a_1 \\ \vdots \\ a_n \end{bmatrix}, \quad \mathbf{V} = \begin{bmatrix} 1 & x_0 & x_0^2 & \cdots & x_0^n \\ 1 & x_1 & x_1^2 & \cdots & x_1^n \\ \vdots & \vdots & \vdots & \ddots & \vdots \\ 1 & x_m & x_m^2 & \cdots & x_m^n \end{bmatrix}$$

Using this notation, Eq. A.5 can be written in matrix form as:

$$\mathbf{V}^t \mathbf{V} \mathbf{a} = \mathbf{V}^t \mathbf{y} \quad \text{(A.6)}$$

where $\mathbf{V}^t\mathbf{V}$ is a $(n+1) \times (n+1)$ symmetric invertible matrix. Thus the solution for the vector $\mathbf{a}$ is:

$$\mathbf{a} = (\mathbf{V}^t\mathbf{V})^{-1}\mathbf{V}^t\mathbf{y} = \mathbf{V}^-\mathbf{y} \qquad (A.7)$$

where $\mathbf{V}^-$ is called the pseudo-inverse of $\mathbf{V}$.

A.1.2 3D interpolation function of the first degree

After this introduction on how to find an interpolation function that minimizes the mean square error in one dimension, we now try to find such a function in 3D to be used for the transformation from the device coordinates given by the scanner to the tristimulus values in CIELAB space. The inputs are thus the three independent variables R, G and B; the outputs are the values $L^{(c)}$, $a^{(c)}$ and $b^{(c)}$.[1] The 3D interpolation function is denoted $f(\cdot)$ as follows:

$$\begin{bmatrix} L^{(c)} & a^{(c)} & b^{(c)} \end{bmatrix} = f\left(\begin{bmatrix} R & G & B \end{bmatrix}\right) \qquad (A.8)$$

To get acquainted with this function, we treat the example of a first order polynomial approximation. This linear transformation can be written as a matrix multiplication as follows:

$$L^{(c)} = \begin{bmatrix} R & G & B \end{bmatrix} \cdot \begin{bmatrix} \alpha_1 \\ \alpha_2 \\ \alpha_3 \end{bmatrix} = R\alpha_1 + G\alpha_2 + B\alpha_3 \qquad (A.9)$$

$$a^{(c)} = \begin{bmatrix} R & G & B \end{bmatrix} \cdot \begin{bmatrix} \beta_1 \\ \beta_2 \\ \beta_3 \end{bmatrix} = R\beta_1 + G\beta_2 + B\beta_3 \qquad (A.10)$$

$$b^{(c)} = \begin{bmatrix} R & G & B \end{bmatrix} \cdot \begin{bmatrix} \gamma_1 \\ \gamma_2 \\ \gamma_3 \end{bmatrix} = R\gamma_1 + G\gamma_2 + B\gamma_3 \qquad (A.11)$$

Our task is to find the optimal set $(\alpha_j, \beta_j, \gamma_j)$, $j = 1, 2, 3$, based on minimizing the mean square error between the calculated values

$$(L_i^{(c)}, a_i^{(c)}, b_i^{(c)})$$

---

[1] In the following we will use the superscript c to denote calculated values. The superscript t denotes theoretical values. We thus omit the asterisk normally associated with the values L^*, a^* and b^*.

and the theoretical ones
$$(L_i^{(t)}, a_i^{(t)}, b_i^{(t)}),$$
corresponding to the $m + 1$ color patches. The subscript i refers to the ith patch. This error is defined in Equation A.12.

$$\text{MSE} = \frac{1}{m+1} \sum_{i=0}^{m} \left((L_i^{(t)} - L_i^{(c)})^2 + (a_i^{(t)} - a_i^{(c)})^2 + (b_i^{(t)} - b_i^{(c)})^2 \right)$$

(A.12)

This equation is an application of the *CIE 1976 L*a*b* color-difference formula* given in Equation 2.29 on page 29. Considering Equation A.12 we see that the three right parts of the equation are all positive. This implies that we can do the minimization separately for each of the L, a and b channels, by differentiating separately with respect to the unknown coefficients.

By replacing $L_i^{(c)}$, $a_i^{(c)}$ and $b_i^{(c)}$ in Equation A.12 by the expressions given in Eqs. A.9 - A.11, we get the following expressions for the mean square errors for each of the channels.

$$\text{MSE}_L = \frac{1}{m+1} \sum_{i=0}^{m} (L_i^{(t)} - (R_i\alpha_1 + G_i\alpha_2 + B_i\alpha_3))^2 \qquad \text{(A.13)}$$

$$\text{MSE}_a = \frac{1}{m+1} \sum_{i=0}^{m} (a_i^{(t)} - (R_i\beta_1 + G_i\beta_2 + B_i\beta_3))^2 \qquad \text{(A.14)}$$

$$\text{MSE}_b = \frac{1}{m+1} \sum_{i=0}^{m} (b_i^{(t)} - (R_i\gamma_1 + G_i\gamma_2 + B_i\gamma_3))^2 \qquad \text{(A.15)}$$

In the following calculations, we will only treat the systems of equations for L, since the calculation for a and b are similar. The mean square error in Equation A.13 is similar to Equation A.3 for the following overdetermined system:

$$L_0^{(t)} = \alpha_0 R_0 + \alpha_1 G_0 + \alpha_2 B_0 \qquad \text{(A.16)}$$
$$L_1^{(t)} = \alpha_0 R_1 + \alpha_1 G_1 + \alpha_2 B_1$$
$$\vdots$$
$$L_m^{(t)} = \alpha_0 R_m + \alpha_1 G_m + \alpha_2 B_m$$

R_i, G_i and B_i are data values playing the same role as the constant 1, x_i and x_i^2, respectively. This system of equations can be written using a matrix form as follows.

$$\begin{bmatrix} L_0^{(t)} \\ L_1^{(t)} \\ \vdots \\ L_{m-1}^{(t)} \end{bmatrix} = \begin{bmatrix} R_0 & G_0 & B_0 \\ R_1 & G_1 & B_1 \\ \vdots & \vdots & \vdots \\ R_m & G_m & B_m \end{bmatrix} \cdot \begin{bmatrix} \alpha_0 \\ \alpha_1 \\ \alpha_2 \end{bmatrix} \qquad \text{(A.17)}$$

or, in a more compact notation,

$$\mathbf{L} = \mathbf{V}\boldsymbol{\alpha} \qquad \text{(A.18)}$$

As developed in Section A.1.1, the optimal solution, found by minimizing the mean square error of this overdetermined system is obtained as follows:

$$\boldsymbol{\alpha} = (\mathbf{V}^t \mathbf{V})^{-1} \mathbf{V}^t \mathbf{L} \qquad \text{(A.19)}$$

We proceed similarly for the vectors $\boldsymbol{\beta}$ and $\boldsymbol{\gamma}$. We thus have expressions to find our coefficients α_j, β_j and γ_i, $j = 1, 2, 3$, which is necessary to perform the polynomial approximation of the first degree of the conversion from RGB space to CIELAB space shown in Eqs. A.9, A.10 and A.11.

We note that Eq. A.19 is equivalent to Eq. A.7 and represents the mean square error solution for the polynomial approximation $f(\cdot)$ of the first degree.

A.1.3   3D interpolation function of general degree n

The method described in the previous section can be extended to an arbitrary degree n.

The equations A.9, A.10 and A.11 extend to

$$L^{(c)} = \mathbf{v}^{(n)t}\boldsymbol{\alpha}, \quad a^{(c)} = \mathbf{v}^{(n)t}\boldsymbol{\beta}, \quad b^{(c)} = \mathbf{v}^{(n)t}\boldsymbol{\gamma} \qquad \text{(A.20)}$$

where $\mathbf{v}^{(n)}$ is a vector, depending on the degree n, that contains all possible products and cross-products of the input R, G and B. We have defined the following $\mathbf{v}^{(n)}$ for the first, second and third degree.

$$
\begin{aligned}
\mathbf{v}^{(1)} &= [1, R, G, B]^t & \text{(A.21)} \\
\mathbf{v}^{(2)} &= [1, R, G, B, R^2, RG, RB, G^2, GB, B^2]^t & \text{(A.22)} \\
\mathbf{v}^{(3)} &= [1, R, G, B, R^2, RG, RB, G^2, GB, B^2, \\
& \quad\ R^3, R^2G, R^2B, RG^2, RGB, \\
& \quad\ RB^2, G^3, G^2B, GB^2, B^3]^t & \text{(A.23)}
\end{aligned}
$$

We remark that the 0'th degree element, that is a constant, is included.

The number $M(n)$ of elements in the vector $\mathbf{v}^{(n)}$ depends on the order n of the 3D interpolation function. We see that we have $M(1) = 4$, $M(2) = 10$, $M(3) = 20$, $M(4) = 35$.

In the following, we will denote $v_i(j)$, the j'th element of the vector $\mathbf{v}_i^{(n)}$ built from the triplet R_i, G_i, B_i. With this notation we extend Equation A.16 to arbitrary order, and get the following set of equations.

$$
L_0^{(t)} = \mathbf{v}_0^{(n)t} \boldsymbol{\alpha} \tag{A.24}
$$

$$
L_1^{(t)} = \mathbf{v}_1^{(n)t} \boldsymbol{\alpha} \tag{A.25}
$$

$$
\vdots \tag{A.26}
$$

$$
L_m^{(t)} = \mathbf{v}_m^{(n)t} \boldsymbol{\alpha} \tag{A.27}
$$

or shorter

$$
\mathbf{L} = \mathbf{V}\boldsymbol{\alpha}, \tag{A.28}
$$

where $\mathbf{V} = [\mathbf{v}_0^{(n)} \mathbf{v}_1^{(n)} \dots \mathbf{v}_m^{(n)}]$ is a $(m + 1) \times M(n)$ matrix. We obtain similarly $\mathbf{a} = \mathbf{V}\boldsymbol{\beta}$ and $\mathbf{b} = \mathbf{V}\boldsymbol{\gamma}$. We can then follow the same steps as for the first order approximation, and get the following solutions for the coefficient vectors $\boldsymbol{\alpha}$, $\boldsymbol{\beta}$ and $\boldsymbol{\gamma}$.

$$
\boldsymbol{\alpha} = (\mathbf{V}^t\mathbf{V})^{-1}\mathbf{V}^t\mathbf{L} \tag{A.29}
$$

$$
\boldsymbol{\beta} = (\mathbf{V}^t\mathbf{V})^{-1}\mathbf{V}^t\mathbf{a} \tag{A.30}
$$

$$
\boldsymbol{\gamma} = (\mathbf{V}^t\mathbf{V})^{-1}\mathbf{V}^t\mathbf{b} \tag{A.31}
$$

A.2 Principal Component Analysis (PCA)

The central goal of the principal component analysis (PCA) is to reduce the dimensionality of a data set which consists of a large number of interrelated variables, while retaining as much as possible of the variation present in the data set. (Jolliffe, 1986, p.1)

Generally we have a set of P observations of N variables.[2] We arrange this in a $(N \times P)$ matrix denoted $\mathbf{R}$, where the columns $\mathbf{r}_j$ are the observations.[3]

$$\mathbf{R} = [\mathbf{r}_1 \mathbf{r}_2 \cdots \mathbf{r}_P] = \begin{bmatrix} r_{11} & r_{12} & \cdots & r_{1P} \\ r_{21} & r_{22} & \cdots & r_{2P} \\ \vdots & \vdots & \ddots & \vdots \\ r_{N1} & r_{N2} & \cdots & r_{NP} \end{bmatrix} \quad (\text{A.32})$$

If the observations $\mathbf{r}_j$ are not centered around their mean, we define the centered matrix of observations as

$$\mathbf{X} = \begin{bmatrix} \mathbf{r}_1 - \bar{\mathbf{r}} & \mathbf{r}_2 - \bar{\mathbf{r}} & \cdots & \mathbf{r}_P - \bar{\mathbf{r}} \end{bmatrix}, \quad (\text{A.33})$$

where $\bar{\mathbf{r}}$ contains the mean values of the observations,

$$\bar{\mathbf{r}} = \frac{1}{P} \sum_{j=1}^{P} \mathbf{r}_j \quad (\text{A.34})$$

We have thus a centered $(N \times P)$ observation matrix $\mathbf{X}$, and we wish to represent each observation using $\tilde{N} < N$ components. The PCA identifies $\tilde{N}$ so-called modes, being defined as the $\tilde{N}$-vectors $\mathbf{u}_j$, $j = 1 \ldots \tilde{N}$ corresponding to the directions in N-dimensional space where the observations exhibit maximum variance. That is, the first mode $\mathbf{u}_1$ corresponds to the direction of maximum variance, $\mathbf{u}_2$ should be the direction of maximum variance, subject to being uncorrelated to $\mathbf{u}_1$, and so on. The set of chosen $\mathbf{u}_j$ defines thus an orthogonal basis of a vector sub-space of dimension $\tilde{N}$.

The representation of an observation $\mathbf{x}$ using its principal components may be expressed as the $\tilde{N}$-vector $\mathbf{z} = [z_1, z_2, \ldots, z_{\tilde{N}}]^t$, where z_j represents the jth principal component (PC) of the observation, expressed by

$$z_j = \mathbf{u}_j^t \mathbf{x} \quad (\text{A.35})$$

Defining the modes matrix $\tilde{\mathbf{U}} = [\mathbf{u}_1 \mathbf{u}_2 \cdots \mathbf{u}_{\tilde{N}}]$, we may reformulate this as a matrix multiplication,

$$\mathbf{z} = \tilde{\mathbf{U}}^t \mathbf{x} \quad (\text{A.36})$$

---

[2] In our work, this corresponds mostly to the spectral reflectances of P samples, quantified on N wavelengths.

[3] Note that in the notation of Jolliffe (1986), the $N \times P$ matrix $\mathbf{R}$ will invariably be a matrix of N observations on P variables. This is exactly the opposite of our notation.

If the representation using PCs is used for the purpose of information compression, it is of great interest to proceed to the reconstruction of an approximation of the original observation, using the PCs. This reconstruction $\tilde{\mathbf{r}}$ can be expressed as a sum of the modes $\mathbf{u}_j$ weighted by the PCs z_j, as follows.

$$\tilde{\mathbf{x}} = \sum_{j=1}^{\tilde{N}} z_j \mathbf{u}_j = \sum_{j=1}^{\tilde{N}} \mathbf{u}_j^t \mathbf{x} \mathbf{u}_j. \tag{A.37}$$

Using matrix notation, Equation A.37 becomes

$$\tilde{\mathbf{x}} = \tilde{\mathbf{U}} \mathbf{z} = \tilde{\mathbf{U}} \tilde{\mathbf{U}}^t \mathbf{x}, \tag{A.38}$$

this representation making it even clearer that the representation using the modes corresponds to a projection of the observation vector $\mathbf{x}$ onto the subspace defined by the modes $\mathbf{u}_j$.

Equation A.38 my also equivalently be represented as

$$\tilde{\mathbf{r}} = \tilde{\mathbf{U}} \tilde{\mathbf{U}}^t (\mathbf{r} - \bar{\mathbf{r}}) + \bar{\mathbf{r}} \tag{A.39}$$

Figure A.1 gives an example of the principal component analysis of 50 observations of the two highly correlated variables x_1 and x_2. If we transform x_1 and x_2 to the PCs z_1 and z_2 we get the plot A.1(c), and we note that the variation is concentrated in z_1.

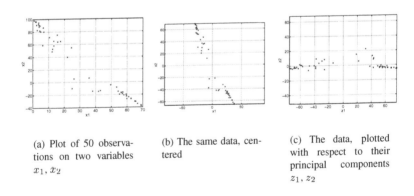

(a) Plot of 50 observations on two variables x_1, x_2

(b) The same data, centered

(c) The data, plotted with respect to their principal components z_1, z_2

Figure A.1: *Illustration of PCA.*

We have not yet discussed the actual determination of the modes for a given data set. An important criterion for this is the reconstruction error

$d = \|\tilde{\mathbf{x}} - \mathbf{x}\|$, that is the distance between the original observation and its projection onto the subspace defined by the modes. We define the mean square error on all the observations $\mathbf{x}_j$ as follows

$$\delta^2 = \sum_{j=1}^{P} \|\tilde{\mathbf{x}} - \mathbf{x}\|^2 \qquad (A.40)$$

It can be shown, (Jolliffe, 1986, p.4) that the optimal jth PC in the sense of minimizing the mean square error δ given above, is given by the eigenvector $\mathbf{u}_j$ of the covariance matrix

$$\mathbf{S} = \frac{1}{P}\mathbf{X}\mathbf{X}^t, \qquad (A.41)$$

corresponding to its jth largest eigenvalue λ_j:

$$\mathbf{S}\mathbf{u}_j = \lambda_j \mathbf{u}_j. \qquad (A.42)$$

Furthermore, if $\mathbf{u}_j$ is chosen to have unit length, then the variance of the mode z_j is given by the eigenvalue λ_j. The diagonalization of the covariance matrix $\mathbf{S}$ does always exist, since $\mathbf{S}$ is symmetric. By denoting as above $\mathbf{U} = [\mathbf{u}_1 \mathbf{u}_2 \cdots \mathbf{u}_N]$, the unitary matrix of the eigenvectors, and

$$\mathbf{\Lambda} = \begin{bmatrix} \lambda_1 & 0 & \cdots & 0 \\ 0 & \lambda_2 & \cdots & 0 \\ \vdots & \vdots & \ddots & \vdots \\ 0 & 0 & \cdots & \lambda_N \end{bmatrix} \qquad (A.43)$$

the diagonal matrix of eigenvalues λ_j arranged in decreasing order, we have the following relation:

$$\mathbf{S}\mathbf{U} = \mathbf{U}\mathbf{\Lambda}, \qquad (A.44)$$

from which we deduce

$$\mathbf{S} = \mathbf{U}\mathbf{\Lambda}\mathbf{U}^t, \qquad (A.45)$$

Note that in the case of $\mathrm{rank}(\mathbf{S}) = R$, $R < N$, then $\lambda_j = 0$ for $R < j \le N$.

A.3   Singular Value Decomposition (SVD)

Numerous variants of the singular value decomposition (SVD) algorithm exist, see *i.e.* Dress *et al.* (1992), Pratt (1978), Golub and Reinsch (1970), Dongarra *et al.* (1979), Jolliffe (1986). We present here two slightly different variants, the differences being mainly the sizes of the involved matrices, the result being equal.

A.3.1   SVD of Jolliffe (1986)

In Chapter 3.5 and Appendix A.1 of Jolliffe (1986) the singular value decomposition is presented as follows. (The notation is modified for consistency.) Given an arbitrary matrix, $\mathbf{X}$, of dimension $N \times P$, $\mathbf{X}$ can be written

$$\mathbf{X} = \mathbf{U}\mathbf{W}\mathbf{V}^t, \tag{A.46}$$

where

1. $\mathbf{U}, \mathbf{V}$ are $(N \times R)$, $(P \times R)$ matrices respectively, each of which has orthonormal columns such that $\mathbf{U}^t\mathbf{U} = \mathbf{I}_R$, $\mathbf{V}^t\mathbf{V} = \mathbf{I}_R$;

2. $\mathbf{W}$ is a $(R \times R)$ diagonal matrix;

3. R is the rank of $\mathbf{X}$.

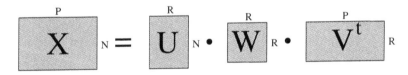

Figure A.2: *Singular values decomposition according to Jolliffe (1986)*

The author presents an important property of the SVD related to PCA, namely that it provides a computationally efficient method of finding the principal components. If we find $\mathbf{U}$, $\mathbf{W}$, $\mathbf{V}$ which satisfy Equation A.46, then $\mathbf{U}$ and $\mathbf{W}$ gives us the eigenvectors and the square roots of the eigenvalues of the matrix $\mathbf{X}\mathbf{X}^t$, and hence the principal components and their variances for the sample covariance matrix $\mathbf{S} = \frac{1}{P}\mathbf{X}\mathbf{X}^t$, as defined in Equation A.41, where $\mathbf{X}$ is measured about its mean.

This representation of the SVD has the advantage to be compact, see Figure A.2, compared to the version we present in the next section, see Figure A.3, but the practical disadvantage that the matrix dimensions are depending on the nature of the data, in particular on the rank R of the observation matrix. In the following we present a slightly different variant of the SVD with predetermined matrix sizes, as found *e.g.* in Pratt (1978), this variant corresponding to what we use in our implementations with Matlab (Matlab Language Reference Manual, 1996).

A.3.2 SVD of Pratt (1978)

It is known (Pratt, 1978, p.126), that for any arbitrary $(N \times P)$ matrix $\mathbf{X}$ of rank R, there exist an $(N \times N)$ unitary matrix $\mathbf{U}$ and a $(P \times P)$ unitary matrix $\mathbf{V}$ for which

$$\mathbf{X} = \mathbf{U}\mathbf{W}\mathbf{V}^t, \qquad (A.47)$$

where $\mathbf{W}$ is an $(N \times P)$ matrix with general diagonal entries w_i, $i = 1 \ldots R$, denoted *singular values* of $\mathbf{X}$. The nature of these matrices is further illustrated in Figures A.3 and A.4.

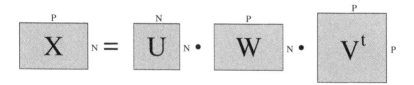

Figure A.3: *Singular values decomposition according to Pratt (1978) for* $N < P$.

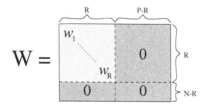

Figure A.4: *The matrix* $\mathbf{W}$ *containing the singular values of* $\mathbf{X}$

The columns of the unitary matrix $\mathbf{U}$ are composed of the eigenvectors $\mathbf{u}_i$, $i = 1 \ldots N$, of the symmetric matrix $\mathbf{XX}^t$. Similarly, the columns of $\mathbf{V}$ are the eigenvectors $\mathbf{v}_i$, $i = 1 \ldots P$, of the symmetric matrix $\mathbf{X}^t\mathbf{X}$.

It is possible to express the matrix decomposition of Equation A.47 in the series form (Pratt, 1978, p.127), this shows us that $\mathbf{X}$ is a weighted sum of the outer product of the eigenvectors $\mathbf{u}_i$ and $\mathbf{v}_i$.

$$\mathbf{X} = \sum_{i=1}^{R} w_i \mathbf{u}_i \mathbf{v}_i^t \qquad (A.48)$$

Note that the SVD may also be presented in a slightly different form, for $N \leq P$, as e.g. in Dress et al. (1992), Golub and Reinsch (1970), where the size of $\mathbf{W}$ is reduced to $(N \times N)$, by eliminating the corresponding zeros off the diagonal, the matrix sizes still being independent of the rank $R \leq N$ of the observation matrix.

A.3.3   Application of the SVD to PCA

As indicated above, the SVD may be used to perform the PCA. We will now relate the matrices involved in the SVD to the PCA.

By multiplying both sides of Equation A.47 by $\mathbf{U}^t$ we obtain the relation

$$\mathbf{U}^t\mathbf{X} = \mathbf{U}^t\mathbf{U}\mathbf{W}\mathbf{V}^t = \mathbf{W}\mathbf{V}^t \qquad (A.49)$$

Applying this, in particular for the ith line and the jth column of the resulting matrix, we have $\mathbf{u}_i^t\mathbf{x}_j = w_i v_{ij}$, where v_{ij} denotes the jth entry of $\mathbf{v}_i$. We can thus express Equation A.48 as follows

$$\mathbf{x}_j = \sum_{i=1}^{R} w_i \mathbf{u}_i v_{ij} = \sum_{i=1}^{R} (\mathbf{u}_i^t \mathbf{x}_j) \mathbf{u}_i, \qquad (A.50)$$

and, using a reduced number $\tilde{N} < R$ of components, we get the approximation $\tilde{\mathbf{x}}_j$ as

$$\tilde{\mathbf{x}}_j = \sum_{i=1}^{\tilde{N}} (\mathbf{u}_i^t \mathbf{x}_j) \mathbf{u}_i, \qquad (A.51)$$

which may be generalized to arbitrary $\mathbf{x}$, giving the same reconstruction formula as employed in PCA, see Equation A.37. Thus the matrix $\mathbf{U}$ contains the modes $\mathbf{u}_j$ of the PCA. This may also be proved by the following result. By combining Equations A.47 and A.41 we have

$$\mathbf{X}\mathbf{X}^t = (\mathbf{U}\mathbf{W}\mathbf{V}^t)(\mathbf{U}\mathbf{W}\mathbf{V}^t)^t = \mathbf{U}\mathbf{W}^2\mathbf{U}^t = \mathbf{U}\Lambda\mathbf{U}^t, \qquad (A.52)$$

Since $\mathbf{S} = \frac{1}{N}\mathbf{X}\mathbf{X}^t = \frac{1}{N}\mathbf{U}\Lambda\mathbf{U}^t$ (*cf.* Equations A.41 and A.45), we have $\Lambda = \frac{1}{N}\mathbf{W}^2$ and the entries on the diagonal of Λ are $\frac{1}{N}w_j^2$, $j = 1 \ldots N$, this showing furthermore that the singular values of $\mathbf{X}$ correspond to the non-negative square roots of the eigenvalues of $\mathbf{X}\mathbf{X}^t$ *i.e.* the variances of the principal components of $\mathbf{X}$.

If the matrix $\mathbf{V}$ is not needed, one might argue that it would be easier to apply the usual diagonalization algorithms to the symmetric matrix $\mathbf{X}\mathbf{X}^t$. However, as pointed out by Golub and Reinsch (1970), the computation of $\mathbf{X}\mathbf{X}^t$ may cause numerical inaccuracy. For example, let

$$\mathbf{X} = \begin{bmatrix} 1 & 0 & \beta \\ 1 & \beta & 0 \end{bmatrix}, \qquad (A.53)$$

then

$$\mathbf{X}\mathbf{X}^t = \begin{bmatrix} 1 + \beta^2 & 1 \\ 1 & 1 + \beta^2 \end{bmatrix}, \qquad (A.54)$$

so that the singular values are given by

$$w_1 = \sqrt{2 + \beta^2}, \quad w_2 = |\beta| \qquad (A.55)$$

If $\beta^2 < \epsilon_0$, the machine precision, the computed $\mathbf{X}\mathbf{X}^t$ has the form

$$\widetilde{\mathbf{X}\mathbf{X}^t} = \begin{bmatrix} 1 & 1 \\ 1 & 1 \end{bmatrix}, \qquad (A.56)$$

and the best one may obtain from diagonalization is

$$\tilde{w}_1 = \sqrt{2}, \quad \tilde{w}_2 = 0. \qquad (A.57)$$

The SVD algorithm avoids the calculation of $\mathbf{X}\mathbf{X}^t$, by first reducing the matrix $\mathbf{X}$ to a bidiagonal form using Householder transformations and then finding the singular values of the bidiagonal matrix using the QR algorithm, as first described by Golub and Reinsch (1970), and used by numerous numerical computing systems (Matlab Language Reference Manual, 1996, Dongarra *et al.*, 1979, Dress *et al.*, 1992).

We conclude thus, as indicated above, that the SVD provides an efficient mean for the realization of the PCA.

A.3.4 Application of the SVD to LMS minimization — pseudoinverse

Let $\mathbf{A}$ be a real $(N \times M)$ matrix. An $(M \times N)$ matrix $\mathbf{B}$ is said to be the pseudoinverse of $\mathbf{A}$ if it satisfies the following four *Moore-Penrose conditions* (see Golub and Reinsch (1970) or Golub and van Loan (1983), p.139):

1. $\mathbf{ABA} = \mathbf{A}$,

2. $\mathbf{BAB} = \mathbf{B}$,

3. $(\mathbf{AB})^t = \mathbf{AB}$,

4. $(\mathbf{BA})^t = \mathbf{BA}$.

The unique solution to the above conditions is denoted by $\mathbf{A}^-$. Remark that these conditions amount to the requirement that $\mathbf{AA}^-$ and $\mathbf{A}^-\mathbf{A}$ be orthogonal projections onto $R(\mathbf{A})$ and $R(\mathbf{A}^t)$, respectively.

It can easily be verified that if $\mathbf{A} = \mathbf{UWV}^t$, cf. Equation A.46, then $\mathbf{A}^- = \mathbf{VW}^-\mathbf{U}^t$, where $\mathbf{W}^- = \mathrm{diag}(w_i^-)$ and

$$w_i^- = \begin{cases} w_i^{-1} & \text{for } w_i > 0 \\ 0 & \text{for } w_i = 0, \end{cases} \tag{A.58}$$

where w_i is the singular values of $\mathbf{A}$. Thus the pseudoinverse may readily be calculated by the SVD.

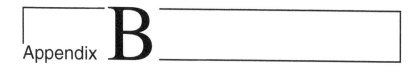

Appendix **B**

Color transformation by 3D interpolation

Introduction

A general color transformation between two color spaces can be described by the following equation:

$$\mathbf{O} = f(\mathbf{P}) \tag{B.1}$$

where $\mathbf{P}$ is the input color signal, and $\mathbf{O}$ denotes the output signal. [1]

For the color transformations needed in our applications, we use a 3D tetrahedron interpolation algorithm proposed by Kanamori *et al.* (1990). The algorithm is able to perform an arbitrary color transform, given the output values for a given subset of the input color space. The two main applications for this algorithm applied to the color facsimile will be the transformation from the RGB-values provided by the scanner to CIELAB-space, as well as from CIELAB- space to the printer device coordinates.

Of course, these transformations could have been performed directly, using the 3rd degree 3D polynomial approximation described in Chapter 3, or using a full 3D look-up table (LUT), but we wish to gain speed, storage space and flexibility by using this algorithm.

---

[1] In our presentation we will often use the conversion RGB $\Rightarrow$ CIELAB as example. This gives $\mathbf{P} = [R\,G\,B]$ and $\mathbf{O} = [L^*\,a^*\,b^*]$. However, the choice of $\mathbf{P}$ and $\mathbf{O}$ is irrelevant to the definition of the algorithm, since it is general.

The speed gain is achieved by a reduced number of operations, as compared to the polynomial approximation. The storage space gain is evident when compared to a conventional LUT. The flexibility stems from the fact that the algorithm is able to perform an arbitrary color transform, or in fact any transform, from one three-dimensional space to another; the performed transform depends uniquely on the choice of output values in the LUT.

We will describe the algorithm by starting with a conventional 3D lookup table and then evolve through conventional 3D interpolation, towards the 3D tetrahedron interpolation algorithm proposed by Kanamori *et al.* (1990).

General transformation using a lookup table

We may represent the general color transformation using a 3D LUT as follows:

$$\mathbf{O_P} = \text{LUT}(i, j, k) = \text{LUT}(\mathbf{P}) \qquad (\text{B.2})$$

where $\mathbf{P} = [i, j, k]$ is the quantized input color signal. The implementation of this three-dimensional LUT requires a very large amount of memory (about 50 Mb when input and output values are stored with 8-bit accuracy).

Transformation using conventional 3D interpolation

To reduce the memory requirements we use a 3D interpolation technique where we store the output values denoted $\mathbf{O}_i$ for a limited number of points $\mathbf{P}_i$ in the input RGB space. We then calculate the output values for any point $\mathbf{P}$ by an interpolation between some of these values.

We divide the RGB - space into a given number of cubic sub-spaces, as shown in Figure B.1 on the next page. With a conventional 3D interpolation method we would calculate the resulting value by first finding the cube in which the input point lies. Then the output values are calculated as an interpolation of the output values $\mathbf{O}_1, \ldots, \mathbf{O}_8$ in the eight corners $\mathbf{P}_1, \ldots, \mathbf{P}_8$, weighted by the linear interpolation coefficients $W_1, \ldots, W_8$. This is indicated in Equation B.3.

$$\mathbf{O_P} = \sum_{i=1}^{8} W_i \mathbf{O}_i \qquad (\text{B.3})$$

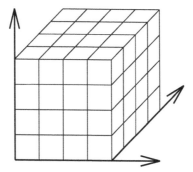

Figure B.1: *Partition of the RGB-space into cubic sub-spaces.*

Transformation using 3D tetrahedral interpolation

It is possible to further reduce by a factor of 2 the computational complexity by introducing tetrahedral sub-spaces. We divide each cubic sub-space into 5 tetrahedral sub-spaces following one of the two schemes outlined in Figure B.2 on the following page. These two subdivisions are just the mirror of each other. We choose each of these two subdivisions for any pair of face-adjacent cubes. The common face will be subdivided in the same two triangles for each of the two cubes. This will guarantee the $\mathbb{C}^0$ continuity of the interpolation scheme when crossing any face shared by two cubes.

We thus calculate the interpolation with 4 multiplications and 3 additions instead of 8 and 7, respectively (Eq. B.3), as follows:

$$\mathbf{O_P} = \sum_{i=1}^{4} \frac{\Delta_i}{\Delta} \mathbf{O}_i \qquad (B.4)$$

where $\mathbf{O}_i$ are the 4 vertices of the tetrahedron containing $\mathbf{P}$, Δ is the volume of the tetrahedron $\mathbf{P}_1\mathbf{P}_2\mathbf{P}_3\mathbf{P}_4$, as shown in Equation B.5, while Δ_i is the volume of the sub-tetrahedron generated by replacing the point $\mathbf{P}_i$ by $\mathbf{P}$, as shown in Equation B.6 and Figure B.3 on page 261.

$$\Delta = \frac{1}{6} \begin{vmatrix} 1 & 1 & 1 & 1 \\ x_1 & x_2 & x_3 & x_4 \\ y_1 & y_2 & y_3 & y_4 \\ z_1 & z_2 & z_3 & z_4 \end{vmatrix} \qquad (B.5)$$

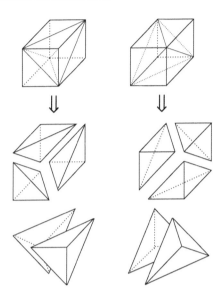

Figure B.2: *The two possible schemes for dividing a cube into five tetra-hedrons. We see that there exists ten different types of tetrahedron.*

$$\Delta_i = \frac{1}{6} \begin{vmatrix} 1 & \overbrace{1}^{\text{col } i} & 1 & 1 \\ x_1 & x & x_3 & x_4 \\ y_1 & y & y_3 & y_4 \\ z_1 & z & z_3 & z_4 \end{vmatrix} \tag{B.6}$$

In Eqs. B.5 and B.6, (x, y, z) and (x_i, y_i, z_i), $i = 1, 2, 3, 4$, denotes the coordinates of P and P_i, respectively.

Thus we obtain an algorithm where the given conversion is calculated as an interpolation between four output values. The operations done offline consist in calculating the output values for all the lattice points, according to the desired transform, and for each point in a cube, the type of the corresponding tetrahedron to which it belongs, and the 4 weighting factors Δ_i/Δ of Equation B.4. The algorithm is shown graphically in Figure B.4.

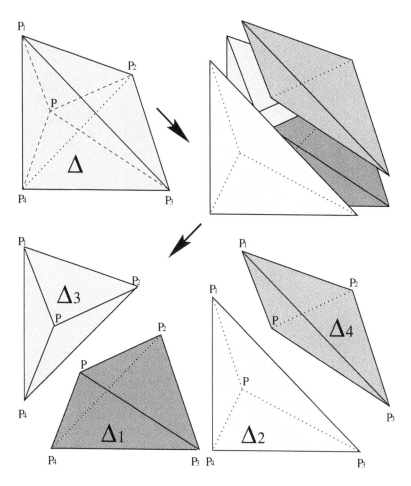

Figure B.3: *The division of a tetrahedron* $P_1 P_2 P_3 P_4$ *into four sub-tetrahedra* $P P_2 P_3 P_4$, $P_1 P P_3 P_4$, $P_1 P_2 P P_4$ *and* $P_1 P_2 P_3 P$, *defined by the input point* P. *We note the graphical interpretation of the volumes* Δ *and* Δ_i *given in Equation B.4.*

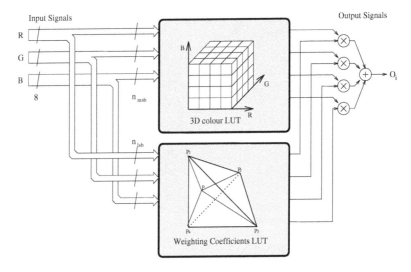

Figure B.4: *3D tetrahedron interpolation color transformation algorithm. The input signals are divided into two parts, the most and least significant bits. The most significant bits are used to find the cubes containing the RGB signal, while the least significant bits determine the tetrahedron containing the signal, as well as the 4 weighting coefficients.*

Appendix C

Scanner characterization data

The tables in this appendix are intended to give more detailed quantitative data for the scanner characterization methods described in Chapter 3. The reported data is obtained by using the IT8.7/2 target from AGFA, and the proposed method labeled (p=1/3, T3, LAB), as described in Section 3.2.3.5.

First, the IT8.7/2 target is scanned with a gamma correction[1] of 1/3, see Figure C.1, and the mean RGB values of each patch are computed, see columns 3-5 of Table C.2. We then apply the characterization algorithm as described in Section 3.2.3.5, using the nominal CIELAB values provided by AGFA (columns 6-8 of Table C.2) as target values for the regression.

We recall that the outcome of the characterization algorithm is a third order polynomial that defines the transformation from the cube-root corrected scanner RGB values to CIELAB values, or more specifically the coefficients α_j, β_j and γ_j, $j = 0 \dots 19$, of the polynomials, as given in Equations C.1-C.3.

---

[1] A gamma correction of 1/3 is achieved by specifying the value of 3 in the scanner's user interface.

$$L^{(c)} = \alpha_0 + \alpha_1 R + \alpha_2 G + \alpha_3 B$$
$$+ \alpha_4 R^2 + \alpha_5 RG + \alpha_6 RB + \alpha_7 G^2 + \alpha_8 GB + \alpha_9 B^2$$
$$+ \alpha_{10} R^3 + \alpha_{11} R^2 G + \alpha_{12} R^2 B + \alpha_{13} RG^2 + \alpha_{14} RGB$$
$$+ \alpha_{15} RB^2 + \alpha_{16} G^3 + \alpha_{17} G^2 B + \alpha_{18} GB^2 + \alpha_{19} B^3 \qquad (C.1)$$

$$a^{(c)} = \beta_0 + \beta_1 R + \beta_2 G + \beta_3 B$$
$$+ \beta_4 R^2 + \beta_5 RG + \beta_6 RB + \beta_7 G^2 + \beta_8 GB + \beta_9 B^2$$
$$+ \beta_{10} R^3 + \beta_{11} R^2 G + \beta_{12} R^2 B + \beta_{13} RG^2 + \beta_{14} RGB$$
$$+ \beta_{15} RB^2 + \beta_{16} G^3 + \beta_{17} G^2 B + \beta_{18} GB^2 + \beta_{19} B^3 \qquad (C.2)$$

$$b^{(c)} = \gamma_0 + \gamma_1 R + \gamma_2 G + \gamma_3 B$$
$$+ \gamma_4 R^2 + \gamma_5 RG + \gamma_6 RB + \gamma_7 G^2 + \gamma_8 GB + \gamma_9 B^2$$
$$+ \gamma_{10} R^3 + \gamma_{11} R^2 G + \gamma_{12} R^2 B + \gamma_{13} RG^2 + \gamma_{14} RGB$$
$$+ \gamma_{15} RB^2 + \gamma_{16} G^3 + \gamma_{17} G^2 B + \gamma_{18} GB^2 + \gamma_{19} B^3 \qquad (C.3)$$

Using the aforementioned equipment and method, we obtain the interpolation coefficients given in Table C.1. Using the resulting polynomial function, we may calculate the resulting CIELAB values of each patch of the target. These values are given in columns 9-11 of Table C.2, and the residual color differences expressed in CIELAB ΔE units are given in the last column. The mean error is 0.918 (*cf.* Table 3.1 on page 63) and a maximal error of 4.666 occur for patch number L19, a saturated blue color.

Table C.1: *Regression polynomial coefficients for the scanner characterization.*

Element	j	α_j	β_j	γ_j
1	0	-1.48E+01	-2.12E+00	3.48E+00
R	1	1.10E-01	6.18E-01	1.39E-01
G	2	2.58E-01	-7.28E-01	6.25E-01
B	3	1.01E-01	1.18E-01	-7.28E-01
RR	4	9.38E-04	-1.45E-03	1.73E-03
RG	5	-1.37E-03	3.88E-03	-2.03E-03
RB	6	-5.56E-04	-9.25E-04	-1.41E-03
GG	7	8.11E-04	1.57E-03	6.65E-04
GB	8	3.03E-04	-7.64E-03	1.61E-03
BB	9	-3.46E-04	4.46E-03	-9.53E-04
RRR	10	-1.71E-06	5.30E-06	-3.90E-06
RRG	11	1.87E-09	-6.29E-06	8.20E-07
RRB	12	8.79E-07	-4.12E-06	4.57E-06
RGG	13	3.04E-06	-1.90E-05	6.59E-06
RGB	14	5.00E-07	3.58E-05	-7.38E-06
RBB	15	-1.85E-07	-1.23E-05	1.51E-06
GGG	16	-2.43E-06	4.19E-06	-2.32E-06
GGB	17	-9.00E-07	7.04E-06	-6.41E-06
GBB	18	3.57E-07	-1.49E-05	8.55E-06
BBB	19	7.11E-07	4.71E-06	-1.35E-06

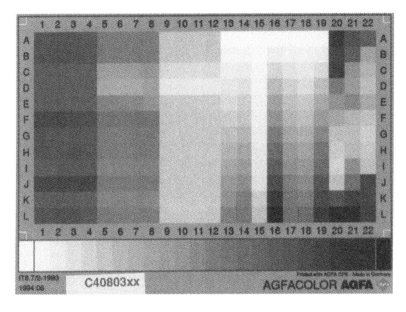

Figure C.1: *The AGFA IT8.7/2 characterization target scanned using an AGFA Arcus 2 flatbed scanner with a gamma correction of 1/3.*

Table C.2: *Scanner characterization data. Average RGB values of all the patches of the AGFA IT7.7.2 color chart, scanned with an AGFA Arcus 2 scanner with a gamma correction of 0.3333, as well as the calculated and nominal CIELAB values. The calculated values are obtained using the third order regression method described in Section 3.2.3.5, and the nominal values are those supplied with the target by AGFA.*

Patch #		Scanned RGB			Calculated CIELAB			Nominal CIELAB			
	i	$R_i^{1/3}$	$G_i^{1/3}$	$B_i^{1/3}$	$L_i^{(c)}$	$a_i^{(c)}$	$b_i^{(c)}$	$L_i^{(t)}$	$a_i^{(t)}$	$b_i^{(t)}$	ΔE_i
A1	1	87.54	66.25	72.08	18.25	11.67	3.03	18.50	11.00	3.23	0.74
A2	2	101.46	61.36	66.71	18.75	22.15	7.38	18.97	21.84	7.48	0.40
A3	3	114.54	55.60	61.69	19.32	32.33	11.43	19.26	32.70	10.52	0.99
A4	4	126.78	56.61	64.23	21.75	38.49	13.05	21.53	39.27	11.78	1.51
A5	5	133.77	112.61	115.22	38.19	11.24	4.73	38.14	11.94	4.31	0.82
A6	6	153.03	106.50	110.35	38.93	26.04	8.79	38.71	27.39	8.16	1.51
A7	7	172.65	98.66	103.59	39.72	41.55	14.41	39.28	43.25	13.65	1.91
A8	8	187.57	92.82	98.65	40.67	53.25	19.09	40.17	54.70	18.02	1.87
A9	9	203.01	187.70	185.42	69.00	7.69	5.13	69.78	7.14	4.81	1.01
A10	10	213.18	185.12	183.67	69.31	15.45	6.56	70.01	15.08	6.18	0.88
A11	11	223.08	181.43	180.93	69.27	23.71	8.16	70.09	23.30	8.18	0.92
A12	12	226.80	177.36	176.09	68.43	28.27	10.32	69.04	28.10	10.49	0.65
A13	13	242.26	241.35	242.47	89.83	0.59	-1.52	90.57	0.33	-1.82	0.84
A14	14	243.00	242.05	243.21	90.12	0.64	-1.56	90.58	0.66	-1.70	0.48
A15	15	243.44	242.38	241.08	90.15	0.44	0.25	90.42	0.39	0.28	0.27
A16	16	243.01	242.20	240.99	90.06	0.28	0.16	90.31	0.36	-0.06	0.34
A17	17	243.63	242.64	241.43	90.26	0.41	0.17	90.36	0.53	0.13	0.16
A18	18	243.33	242.80	241.46	90.27	0.12	0.19	90.35	0.22	0.31	0.17
A19	19	242.63	241.98	242.84	90.05	0.42	-1.38	90.38	0.58	-1.32	0.37
A20	20	71.41	49.43	53.22	10.82	11.69	5.01	8.75	14.45	4.15	3.55
A21	21	118.08	54.60	50.57	19.35	33.97	21.31	19.85	34.16	21.85	0.76
A22	22	155.03	59.38	49.61	27.22	51.80	35.24	28.07	51.20	36.55	1.67
B1	23	84.37	66.85	67.86	17.82	8.79	6.35	18.53	8.10	7.11	1.25
B2	24	96.68	64.23	59.90	18.49	16.86	13.81	18.95	16.69	14.36	0.74
B3	25	106.61	60.40	52.78	18.80	24.39	19.78	19.21	24.51	20.50	0.84
B4	26	113.88	60.76	52.96	20.09	28.23	21.66	20.31	28.66	22.74	1.18
B5	27	135.59	111.72	106.83	37.95	11.96	11.63	37.88	12.72	11.54	0.77
B6	28	153.16	107.23	93.73	38.79	24.33	23.58	38.50	25.38	23.42	1.10

continued on next page

continued from previous page

	i	$R_i^{1/3}$	$G_i^{1/3}$	$B_i^{1/3}$	$L_i^{(c)}$	$a_i^{(c)}$	$b_i^{(c)}$	$L_i^{(t)}$	$a_i^{(t)}$	$b_i^{(t)}$	ΔE_i
B7	29	174.21	99.88	78.30	39.83	40.89	37.49	39.26	41.81	37.13	1.14
B8	30	186.26	94.95	70.56	40.53	51.11	44.46	39.90	51.09	44.16	0.69
B9	31	204.07	187.76	180.64	69.01	7.77	9.24	69.28	7.07	8.73	0.90
B10	32	213.07	186.50	174.52	69.48	13.55	14.90	69.60	12.84	14.72	0.75
B11	33	225.46	185.21	170.14	70.37	21.67	19.77	70.55	20.94	19.78	0.75
B12	34	229.29	182.32	165.09	69.89	25.62	22.88	69.79	24.99	23.15	0.69
B13	35	231.38	239.25	242.45	88.18	-4.58	-4.15	88.12	-4.95	-4.61	0.60
B14	36	243.66	236.39	241.65	88.59	4.80	-3.60	88.27	4.93	-4.03	0.55
B15	37	244.89	241.68	225.51	89.46	0.34	11.70	89.27	-0.08	11.86	0.49
B16	38	229.38	229.98	228.87	84.89	-0.95	0.31	84.64	-1.07	0.14	0.33
B17	39	242.89	232.68	222.44	86.73	4.79	8.64	86.51	4.76	8.86	0.31
B18	40	230.27	237.20	223.79	86.73	-5.79	8.47	86.53	-6.07	8.47	0.34
B19	41	228.16	230.04	238.13	85.17	-0.54	-6.79	85.09	-0.20	-7.03	0.42
B20	42	94.94	55.29	54.49	15.79	21.15	12.56	14.86	21.93	13.77	1.72
B21	43	154.31	62.86	52.44	27.81	49.59	34.38	28.11	49.25	35.84	1.53
B22	44	191.73	150.90	139.67	56.45	21.79	18.63	57.34	21.98	18.57	0.91
C1	45	92.00	82.56	77.41	23.49	3.28	9.89	23.84	3.10	9.98	0.40
C2	46	97.47	82.20	68.41	23.69	6.21	18.05	23.95	6.43	17.95	0.35
C3	47	105.29	82.20	59.57	24.35	10.79	26.89	24.67	11.53	26.63	0.84
C4	48	110.63	83.46	58.67	25.40	13.25	29.54	25.51	14.04	28.80	1.09
C5	49	168.96	151.11	131.08	53.62	7.43	21.92	53.70	7.69	21.68	0.37
C6	50	184.41	148.95	110.22	54.46	17.97	41.14	54.27	17.81	41.36	0.33
C7	51	196.06	145.56	86.02	54.53	28.25	62.34	54.21	27.33	62.40	0.98
C8	52	198.42	132.24	72.25	51.13	38.26	66.97	50.43	36.91	65.95	1.83
C9	53	215.63	205.45	190.24	75.45	3.65	13.97	75.39	3.28	13.28	0.78
C10	54	226.39	206.52	179.23	76.59	8.86	25.14	76.31	8.43	24.81	0.61
C11	55	233.47	205.32	169.43	76.78	13.58	33.58	76.66	13.02	33.15	0.72
C12	56	235.66	203.14	159.37	76.20	16.06	41.02	75.97	15.60	40.73	0.58
C13	57	210.64	232.22	239.47	84.10	-12.70	-8.63	83.76	-13.11	-9.39	0.93
C14	58	240.66	219.81	234.14	83.39	13.88	-8.22	82.86	14.22	-8.99	1.00
C15	59	244.82	239.71	211.18	88.41	0.63	21.57	88.13	0.12	21.92	0.68
C16	60	204.95	203.17	201.29	74.03	-0.24	2.23	73.59	-0.17	1.96	0.52
C17	61	238.29	212.02	202.12	79.92	14.06	11.77	79.81	13.85	12.04	0.35
C18	62	207.98	226.97	207.14	81.10	-14.02	12.34	81.04	-14.50	12.64	0.57
C19	63	204.07	207.06	226.95	75.93	0.41	-15.56	75.83	0.93	-16.18	0.81
C20	64	75.62	55.64	55.52	13.12	10.06	7.81	12.20	10.91	9.08	1.78

continued on next page

continued from previous page

	i	$R_i^{1/3}$	$G_i^{1/3}$	$B_i^{1/3}$	$L_i^{(c)}$	$a_i^{(c)}$	$b_i^{(c)}$	$L_i^{(t)}$	$a_i^{(t)}$	$b_i^{(t)}$	ΔE_i
C21	65	186.89	128.36	106.43	49.25	31.79	32.57	49.33	32.54	32.81	0.79
C22	66	198.81	170.90	159.07	63.21	14.17	15.82	63.95	14.30	15.45	0.84
D1	67	86.76	85.64	78.46	23.81	-2.02	10.20	24.22	-2.00	10.45	0.48
D2	68	88.39	86.91	70.22	23.98	-2.37	18.00	24.50	-2.05	18.12	0.62
D3	69	91.37	90.72	63.25	25.08	-2.91	26.58	25.24	-2.12	26.24	0.88
D4	70	95.25	91.59	62.07	25.73	-0.97	28.76	25.68	-0.15	27.80	1.26
D5	71	168.56	167.24	142.47	58.43	-2.82	22.43	58.58	-2.68	22.43	0.20
D6	72	170.28	170.12	124.24	58.98	-3.40	39.63	58.83	-3.77	39.88	0.47
D7	73	170.68	172.99	101.25	59.18	-2.91	60.80	58.87	-3.99	61.19	1.19
D8	74	168.30	162.52	82.46	55.40	3.37	69.96	54.60	2.44	69.41	1.35
D9	75	223.58	223.68	202.92	81.67	-2.58	15.83	81.50	-2.64	15.44	0.43
D10	76	228.10	226.01	191.10	82.41	-1.86	27.17	82.07	-1.86	26.82	0.49
D11	77	228.40	226.95	179.28	82.36	-2.49	37.26	82.02	-2.39	36.95	0.47
D12	78	230.47	227.06	169.36	82.35	-1.37	45.75	81.85	-1.19	45.54	0.57
D13	79	193.68	225.56	236.46	80.52	-18.83	-12.26	80.15	-18.88	-12.99	0.82
D14	80	236.33	202.59	225.58	77.77	22.75	-12.79	77.24	23.23	-13.58	1.07
D15	81	244.20	237.17	196.24	87.19	1.19	31.78	86.85	0.86	32.16	0.61
D16	82	183.65	181.07	178.42	64.90	-0.22	3.84	64.40	-0.18	3.39	0.68
D17	83	232.73	192.36	180.94	73.35	22.14	16.33	73.33	21.87	16.29	0.27
D18	84	189.48	217.06	190.56	75.98	-20.05	17.04	76.01	-20.74	17.18	0.70
D19	85	181.80	185.06	214.91	66.98	2.53	-23.24	67.11	2.71	-23.86	0.65
D20	86	157.94	110.21	90.77	40.18	25.26	28.96	39.73	26.72	29.67	1.68
D21	87	166.30	150.88	141.50	53.48	6.63	12.58	53.70	6.82	12.15	0.52
D22	88	203.96	184.69	168.67	67.86	8.79	17.15	68.42	8.79	17.31	0.58
E1	89	81.21	87.35	79.90	23.78	-6.62	9.37	24.01	-6.46	9.57	0.34
E2	90	75.78	91.02	72.62	23.97	-13.16	16.98	24.22	-12.32	17.18	0.90
E3	91	73.76	96.22	66.11	24.99	-17.86	25.33	25.28	-16.25	25.05	1.66
E4	92	76.35	96.99	65.81	25.45	-16.60	26.40	25.39	-15.27	26.22	1.34
E5	93	123.78	134.54	120.12	43.57	-10.08	13.42	43.54	-9.60	13.39	0.48
E6	94	119.02	138.63	105.92	43.82	-16.12	27.05	43.57	-16.27	27.47	0.51
E7	95	111.55	146.12	93.17	44.82	-24.41	41.17	44.54	-25.13	42.09	1.20
E8	96	107.32	158.51	89.95	47.99	-33.02	51.13	47.43	-34.58	51.44	1.69
E9	97	188.38	196.39	179.57	69.79	-7.84	13.12	69.50	-7.30	12.93	0.64
E10	98	188.28	202.09	174.07	71.22	-11.88	20.99	70.89	-11.52	20.38	0.78
E11	99	184.02	205.09	166.18	71.39	-16.47	28.53	70.92	-16.40	28.03	0.69
E12	100	179.99	206.95	159.10	71.28	-19.89	34.77	70.78	-20.03	34.41	0.63

continued on next page

continued from previous page

	i	$R_i^{1/3}$	$G_i^{1/3}$	$B_i^{1/3}$	$L_i^{(c)}$	$a_i^{(c)}$	$b_i^{(c)}$	$L_i^{(t)}$	$a_i^{(t)}$	$b_i^{(t)}$	ΔE_i
E13	101	171.75	215.24	231.68	75.48	-25.66	-17.08	75.47	-25.40	-17.32	0.35
E14	102	231.27	186.49	216.71	72.38	30.45	-16.63	72.02	30.98	-17.24	0.88
E15	103	243.11	233.79	180.53	85.75	2.18	42.36	85.48	2.19	42.75	0.47
E16	104	162.07	160.98	160.09	56.38	-1.13	2.95	55.98	-0.97	2.40	0.70
E17	105	226.93	175.10	163.30	67.61	28.81	19.66	67.60	28.60	19.80	0.26
E18	106	167.08	204.71	173.86	69.83	-26.96	20.07	70.01	-27.73	19.90	0.81
E19	107	160.43	165.95	203.14	58.92	3.57	-29.31	59.20	3.50	-29.91	0.67
E20	108	200.61	121.29	68.62	48.71	45.97	64.27	48.69	44.70	64.62	1.31
E21	109	166.30	126.02	89.68	45.46	21.03	41.18	45.69	21.22	41.10	0.31
E22	110	195.19	156.78	126.53	58.26	19.66	34.03	58.57	19.70	34.23	0.37
F1	111	55.12	66.01	68.46	14.38	-8.42	1.04	14.75	-8.90	1.68	0.88
F2	112	47.40	70.67	66.38	15.02	-17.55	5.18	15.50	-18.89	5.88	1.59
F3	113	44.63	75.99	66.01	16.44	-23.69	8.98	16.69	-25.55	9.52	1.95
F4	114	45.28	79.51	67.96	17.72	-25.82	9.98	17.74	-27.54	10.17	1.73
F5	115	95.19	114.52	111.14	34.55	-14.67	4.00	34.28	-14.26	3.59	0.64
F6	116	82.33	119.97	108.35	35.11	-27.94	8.94	34.84	-27.41	8.88	0.60
F7	117	66.56	122.73	102.75	34.76	-41.79	14.54	34.20	-39.91	14.48	1.97
F8	118	53.61	127.44	100.52	35.65	-55.39	19.53	34.94	-53.61	19.32	1.93
F9	119	185.75	196.62	188.16	69.84	-8.92	6.04	69.92	-8.14	5.90	0.80
F10	120	178.68	199.26	187.09	69.85	-15.32	7.65	69.76	-14.53	7.55	0.80
F11	121	172.78	203.49	187.90	70.50	-21.94	8.97	70.40	-21.25	8.74	0.74
F12	122	167.31	205.21	185.56	70.39	-26.81	11.27	70.14	-26.40	11.23	0.48
F13	123	157.98	208.13	228.29	72.20	-29.51	-20.04	72.17	-29.46	-20.04	0.06
F14	124	227.51	175.68	210.44	68.72	35.39	-19.01	68.61	35.78	-19.36	0.54
F15	125	242.27	231.30	170.32	84.74	3.04	49.21	84.54	3.28	49.56	0.47
F16	126	146.73	146.73	146.46	50.27	-1.81	2.78	49.90	-1.81	2.02	0.85
F17	127	222.69	163.89	150.68	63.86	32.79	22.84	63.87	32.59	23.23	0.44
F18	128	152.61	196.07	162.36	65.67	-31.04	22.27	65.83	-32.13	21.79	1.20
F19	129	145.27	153.19	195.05	53.45	4.04	-33.31	53.42	3.87	-34.27	0.97
F20	130	179.50	162.60	143.17	58.35	6.92	20.73	58.40	7.31	21.22	0.63
F21	131	201.82	172.76	138.49	63.67	13.96	34.75	63.61	13.88	34.70	0.11
F22	132	205.79	183.50	156.17	67.46	10.04	27.08	67.63	9.85	27.03	0.26
G1	133	60.90	76.16	83.52	18.73	-10.34	-3.49	18.93	-10.54	-2.71	0.83
G2	134	51.56	80.79	87.74	19.69	-19.78	-4.23	19.69	-20.14	-3.84	0.53
G3	135	45.63	84.13	91.29	20.61	-26.22	-4.81	20.48	-28.01	-4.49	1.83
G4	136	45.98	88.74	95.64	22.38	-29.31	-4.81	22.10	-30.52	-4.71	1.24

continued on next page

	i	$R_i^{1/3}$	$G_i^{1/3}$	$B_i^{1/3}$	$L_i^{(c)}$	$a_i^{(c)}$	$b_i^{(c)}$	$L_i^{(t)}$	$a_i^{(t)}$	$b_i^{(t)}$	ΔE_i
						continued from previous page					
G5	137	103.68	124.88	126.89	39.04	-15.07	-0.88	39.01	-14.82	-1.23	0.43
G6	138	91.09	128.69	130.54	39.41	-26.16	-2.07	39.35	-25.56	-2.43	0.70
G7	139	72.42	130.90	133.16	39.22	-41.22	-2.93	39.01	-38.86	-3.72	2.50
G8	140	57.05	133.14	138.13	39.78	-54.53	-4.44	39.35	-51.71	-5.38	3.00
G9	141	183.19	196.54	192.61	69.70	-9.99	2.14	69.79	-9.38	2.66	0.81
G10	142	173.87	200.20	196.03	69.96	-18.17	0.68	69.81	-17.30	1.01	0.94
G11	143	164.88	202.38	198.79	69.86	-25.31	-0.95	69.88	-24.44	-0.52	0.97
G12	144	160.54	204.94	202.18	70.34	-29.72	-2.28	70.27	-29.10	-1.83	0.77
G13	145	137.02	196.46	222.27	67.06	-34.81	-24.18	67.22	-34.88	-23.91	0.33
G14	146	221.35	158.89	200.16	63.07	42.76	-22.29	63.06	43.14	-22.43	0.40
G15	147	240.88	226.70	152.93	82.96	4.91	60.97	82.88	5.42	60.91	0.52
G16	148	127.24	127.43	127.49	42.05	-1.96	3.09	41.42	-1.74	2.32	1.02
G17	149	216.00	147.25	131.79	58.36	38.48	27.83	58.34	38.59	27.91	0.14
G18	150	135.15	184.94	146.83	60.41	-35.48	25.84	60.70	-36.79	25.71	1.34
G19	151	125.98	134.85	182.98	45.80	6.54	-38.48	45.91	6.17	-39.44	1.04
G20	152	201.02	185.93	153.43	67.55	5.54	30.03	67.94	5.49	30.12	0.40
G21	153	200.28	183.49	136.30	66.42	6.57	42.78	66.69	6.30	42.63	0.41
G22	154	219.55	199.92	101.06	72.26	10.13	86.73	72.67	10.81	86.62	0.80
H1	155	63.01	74.25	88.12	18.56	-6.30	-8.38	18.96	-6.80	-7.53	1.07
H2	156	55.28	78.58	99.76	19.87	-12.06	-15.34	20.13	-12.17	-14.72	0.69
H3	157	45.56	77.76	105.49	19.34	-15.75	-20.94	19.37	-17.44	-20.25	1.82
H4	158	46.12	83.15	112.23	21.46	-18.57	-22.26	21.33	-19.68	-22.14	1.12
H5	159	107.42	123.37	134.37	39.16	-9.96	-7.72	39.12	-9.82	-8.01	0.33
H6	160	95.56	126.67	145.67	39.68	-17.93	-15.49	39.58	-17.42	-16.07	0.78
H7	161	79.10	127.17	156.67	39.40	-26.27	-24.34	39.22	-24.59	-25.27	1.93
H8	162	65.23	127.43	166.90	39.56	-32.49	-31.73	39.23	-30.23	-32.88	2.56
H9	163	182.36	194.79	198.58	69.31	-8.42	-3.80	69.35	-7.74	-3.09	0.98
H10	164	176.97	197.43	205.73	69.80	-12.70	-8.34	69.79	-12.01	-7.57	1.04
H11	165	170.33	200.22	216.88	70.43	-17.07	-15.82	70.37	-16.28	-15.29	0.95
H12	166	164.14	202.27	223.98	70.81	-21.25	-20.37	70.77	-20.80	-19.85	0.68
H13	167	121.04	187.04	217.29	63.09	-38.41	-27.16	63.34	-38.55	-26.65	0.58
H14	168	214.28	141.31	188.74	57.19	49.92	-25.16	57.43	50.16	-25.01	0.37
H15	169	239.56	223.13	141.37	81.61	6.34	68.61	81.58	7.17	68.79	0.85
H16	170	109.88	109.87	111.83	34.59	-1.60	2.01	33.62	-1.54	1.18	1.28
H17	171	209.26	131.67	116.37	53.30	43.56	30.66	53.06	44.14	31.01	0.71
H18	172	118.33	174.60	134.68	55.57	-40.13	27.36	55.62	-41.53	27.45	1.40

continued on next page

continued from previous page

	i	$R_i^{1/3}$	$G_i^{1/3}$	$B_i^{1/3}$	$L_i^{(c)}$	$a_i^{(c)}$	$b_i^{(c)}$	$L_i^{(t)}$	$a_i^{(t)}$	$b_i^{(t)}$	ΔE_i
H19	173	107.65	116.12	170.44	38.12	10.37	-43.48	38.04	10.02	-44.54	1.12
H20	174	204.94	199.11	155.67	71.70	-0.10	36.71	72.33	-0.16	36.70	0.64
H21	175	183.15	182.25	166.34	64.88	-2.49	14.35	65.51	-2.20	13.66	0.98
H22	176	190.02	197.75	141.66	69.34	-7.93	45.10	69.93	-8.29	45.10	0.69
I1	177	80.78	84.50	104.63	23.94	0.02	-13.03	23.65	-0.36	-13.07	0.47
I2	178	74.59	83.85	116.12	23.65	0.68	-23.79	23.73	0.58	-23.72	0.14
I3	179	66.89	84.43	129.46	23.79	1.24	-35.19	23.68	2.12	-35.26	0.89
I4	180	58.13	84.94	141.52	23.98	1.95	-45.33	23.93	3.57	-45.43	1.62
I5	181	126.40	130.57	143.52	43.36	-2.14	-8.27	43.53	-1.90	-8.00	0.40
I6	182	122.74	130.21	154.34	43.22	-1.35	-17.99	43.31	-1.38	-17.90	0.13
I7	183	119.67	130.90	165.88	43.52	-0.11	-27.49	43.29	-0.46	-27.50	0.42
I8	184	117.77	132.64	177.32	44.28	1.53	-36.00	44.08	0.80	-36.03	0.76
I9	185	191.76	193.32	197.89	69.79	-1.60	-3.00	69.86	-1.32	-2.20	0.85
I10	186	191.13	194.34	205.97	70.28	-1.45	-8.86	70.14	-1.24	-7.92	0.97
I11	187	188.66	193.64	213.96	70.10	-0.81	-15.96	70.02	-0.42	-15.36	0.72
I12	188	188.93	195.22	221.24	70.85	-0.28	-20.63	70.60	0.10	-20.08	0.71
I13	189	106.88	177.79	211.92	59.35	-40.92	-29.53	59.62	-41.45	-29.10	0.73
I14	190	209.23	129.42	180.34	53.26	54.34	-26.45	53.48	54.66	-26.39	0.39
I15	191	238.53	220.11	132.38	80.49	7.65	74.50	80.52	8.60	74.74	0.98
I16	192	96.87	96.05	99.49	28.75	-0.87	1.21	27.56	-1.15	0.48	1.43
I17	193	203.28	118.83	102.47	49.13	47.42	34.06	48.84	48.12	34.82	1.08
I18	194	103.60	164.76	122.32	51.04	-43.67	29.57	50.99	-45.09	29.80	1.44
I19	195	93.66	103.30	161.65	32.76	12.43	-46.83	32.52	12.32	-47.78	0.98
I20	196	198.32	208.21	157.59	73.52	-9.42	39.46	74.22	-9.58	39.82	0.80
I21	197	116.15	150.43	100.97	46.85	-24.65	38.27	46.86	-25.59	38.80	1.08
I22	198	156.83	194.10	158.91	65.40	-27.02	24.23	66.39	-27.47	23.64	1.24
J1	199	63.42	57.71	81.46	13.44	6.79	-14.35	13.86	6.17	-14.10	0.79
J2	200	63.96	56.50	94.20	13.70	12.90	-25.97	13.93	12.51	-25.72	0.52
J3	201	64.31	57.58	109.64	14.65	19.42	-38.62	14.69	19.52	-38.20	0.44
J4	202	61.27	56.24	120.37	14 37	25.83	-49.52	14.51	26.24	-48.63	0.99
J5	203	119.49	116.98	133.04	38.33	2.65	-9.67	38.45	2.52	-9.68	0.18
J6	204	121.70	114.76	141.76	38.13	8.21	-18.32	38.36	7.73	-18.54	0.57
J7	205	122.81	113.43	153.49	38.15	14.10	-29.23	38.20	13.15	-29.05	0.97
J8	206	124.80	113.85	161.49	38.68	18.07	-35.64	38.52	16.77	-35.31	1.35
J9	207	194.60	191.78	197.97	69.64	1.29	-3.66	70.04	1.12	-2.52	1.22
J10	208	197.07	190.58	204.24	69.73	4.76	-9.11	69.74	4.55	-8.48	0.66

continued on next page

continued from previous page											
	i	$R_i^{1/3}$	$G_i^{1/3}$	$B_i^{1/3}$	$L_i^{(c)}$	$a_i^{(c)}$	$b_i^{(c)}$	$L_i^{(t)}$	$a_i^{(t)}$	$b_i^{(t)}$	ΔE_i
J11	209	200.00	191.44	214.34	70.59	7.85	-16.23	70.45	7.74	-15.54	0.72
J12	210	200.55	189.84	218.68	70.32	10.49	-20.68	70.15	10.27	-20.11	0.63
J13	211	86.63	164.05	203.44	53.97	-44.07	-32.38	54.09	-44.49	-32.27	0.45
J14	212	200.70	110.84	166.87	47.22	60.62	-28.06	47.37	60.90	-27.92	0.35
J15	213	236.33	214.94	118.46	78.56	9.76	83.35	78.75	10.95	83.54	1.22
J16	214	74.76	72.71	77.68	18.75	0.13	0.52	18.61	-0.99	0.12	1.20
J17	215	194.93	102.08	82.69	43.71	52.08	39.96	43.64	52.50	40.15	0.46
J18	216	86.32	152.71	107.31	45.55	-47.46	32.21	45.74	-48.19	32.48	0.80
J19	217	71.56	80.76	146.00	23.58	17.95	-52.31	23.60	18.35	-52.31	0.40
J20	218	159.02	197.24	173.14	66.96	-27.46	15.19	68.21	-27.77	15.39	1.31
J21	219	59.45	124.60	76.22	33.46	-46.56	35.58	33.58	-44.51	35.09	2.11
J22	220	45.09	72.10	60.35	14.91	-20.80	10.67	15.26	-23.87	11.67	3.25
K1	221	93.92	75.67	97.86	22.55	13.48	-11.56	23.06	11.98	-11.26	1.62
K2	222	106.24	70.92	106.41	23.05	26.80	-20.00	23.23	24.88	-19.43	2.01
K3	223	117.90	66.98	116.94	23.86	39.96	-29.70	24.03	37.87	-29.14	2.18
K4	224	126.25	65.36	125.93	24.80	49.15	-37.19	25.03	46.48	-36.39	2.80
K5	225	138.15	125.42	141.58	43.00	8.74	-8.54	43.30	8.66	-8.50	0.31
K6	226	148.73	119.88	148.25	42.75	20.99	-16.29	42.98	20.90	-16.30	0.25
K7	227	160.80	117.25	158.19	43.62	32.69	-24.59	43.75	32.39	-24.60	0.33
K8	228	175.50	114.95	169.55	44.98	46.03	-33.36	45.09	45.24	-32.70	1.04
K9	229	200.80	189.14	197.17	69.49	6.90	-3.81	69.70	6.24	-3.08	1.01
K10	230	209.58	187.50	203.15	70.09	14.38	-8.38	69.95	13.96	-7.88	0.67
K11	231	217.30	185.63	209.94	70.52	21.66	-13.90	70.43	21.28	-13.25	0.76
K12	232	222.97	184.98	214.89	71.05	26.48	-17.44	70.99	26.09	-17.07	0.54
K13	233	76.13	155.57	197.83	50.75	-44.49	-33.90	50.94	-45.38	-33.81	0.92
K14	234	192.54	95.27	154.33	42.20	64.76	-28.25	42.18	64.99	-28.22	0.23
K15	235	228.96	195.58	92.47	71.96	18.16	93.60	72.21	18.66	93.48	0.57
K16	236	60.43	57.21	62.13	12.03	0.82	0.88	11.79	-0.13	0.00	1.32
K17	237	186.70	86.47	65.56	38.72	56.09	44.36	39.11	55.40	43.72	1.02
K18	238	71.44	141.75	93.77	40.58	-50.28	34.51	41.01	-49.62	34.87	0.87
K19	239	57.77	64.15	133.09	16.98	22.87	-54.94	17.25	23.99	-54.46	1.25
K20	240	62.80	80.72	77.62	19.97	-13.79	4.61	20.94	-13.59	5.84	1.59
K21	241	49.13	63.98	70.22	13.35	-10.31	-2.37	14.36	-12.34	-1.07	2.61
K22	242	50.33	68.34	88.08	15.75	-9.03	-13.59	16.40	-10.85	-12.70	2.13
L1	243	87.09	66.51	80.09	18.57	12.69	-3.63	18.71	11.51	-3.29	1.24
L2	244	99.63	60.91	80.90	18.84	24.06	-5.47	18.99	22.72	-5.58	1.35
										continued on next page	

continued from previous page

	i	$R_i^{1/3}$	$G_i^{1/3}$	$B_i^{1/3}$	$L_i^{(c)}$	$a_i^{(c)}$	$b_i^{(c)}$	$L_i^{(t)}$	$a_i^{(t)}$	$b_i^{(t)}$	ΔE_i
L3	245	111.42	54.54	82.03	19.13	35.15	-7.70	19.07	34.53	-8.63	1.12
L4	246	122.27	56.71	87.78	21.54	40.39	-8.80	21.15	39.98	-9.68	1.05
L5	247	133.92	111.86	124.55	38.22	13.45	-3.69	38.30	13.60	-3.79	0.20
L6	248	153.73	106.96	126.80	39.43	28.79	-5.09	39.33	29.43	-5.46	0.75
L7	249	174.06	98.39	127.47	40.11	46.09	-6.52	39.87	46.86	-7.20	1.06
L8	250	183.54	94.59	128.34	40.63	53.90	-7.29	40.10	54.78	-7.72	1.12
L9	251	203.74	187.81	191.87	69.27	8.86	0.08	69.56	8.02	0.86	1.18
L10	252	212.66	186.60	193.71	69.90	15.37	-0.81	69.82	14.88	-0.50	0.58
L11	253	224.75	184.01	194.90	70.46	24.63	-1.47	70.46	24.04	-0.92	0.81
L12	254	228.30	179.10	194.42	69.43	30.25	-3.53	69.66	29.60	-3.02	0.85
L13	255	59.83	137.53	185.25	43.94	-40.45	-37.22	44.04	-44.39	-36.46	4.01
L14	256	186.52	85.81	145.99	39.08	66.44	-27.85	38.94	66.65	-27.91	0.26
L15	257	223.93	180.42	85.90	67.19	24.76	89.21	67.39	24.17	89.20	0.62
L16	258	50.71	47.21	51.17	7.55	0.82	1.72	6.36	0.63	-0.65	2.66
L17	259	177.77	73.02	52.76	34.20	58.22	46.04	34.78	56.56	45.77	1.78
L18	260	53.15	116.59	73.37	30.39	-46.10	32.01	30.67	-45.64	31.32	0.87
L19	261	47.72	51.62	120.21	11.93	24.54	-54.02	11.62	29.13	-54.81	4.67
L20	262	44.71	48.89	58.43	7.83	-2.88	-3.77	7.20	-3.30	-4.37	0.97
L21	263	51.98	47.18	58.09	8.08	2.89	-3.70	7.86	4.41	-5.14	2.10
L22	264	75.97	54.18	64.94	13.22	12.76	-0.86	13.67	13.24	-0.77	0.66
D_{min}	265	242.96	242.60	243.95	90.30	0.32	-1.80	90.70	0.59	-1.75	0.49
1	266	241.95	241.35	239.19	89.65	0.03	0.86	89.77	0.14	0.87	0.16
2	267	230.76	230.83	227.78	85.22	-0.82	1.83	85.19	-0.72	2.12	0.31
3	268	215.48	217.73	214.74	79.61	-2.51	1.99	79.44	-2.23	2.04	0.33
4	269	205.30	203.90	203.50	74.34	-0.28	0.98	74.05	-0.06	1.20	0.42
5	270	194.83	193.62	193.80	70.07	-0.52	0.88	69.89	-0.34	1.05	0.30
6	271	182.70	180.84	182.13	64.84	-0.15	0.54	64.71	0.05	0.62	0.25
7	272	172.37	171.65	173.23	60.88	-0.92	0.48	60.68	-0.70	0.64	0.34
8	273	162.92	161.94	164.00	56.85	-0.78	0.46	56.73	-0.46	0.48	0.34
9	274	152.83	152.85	154.98	52.92	-1.45	0.58	52.84	-1.18	0.74	0.32
10	275	145.46	146.41	148.73	50.10	-2.03	0.51	49.97	-1.93	0.82	0.35
11	276	136.39	136.56	138.56	46.00	-1.64	1.20	45.71	-1.44	1.24	0.36
12	277	126.27	126.70	128.83	41.78	-1.82	1.36	41.37	-1.67	1.42	0.44
13	278	116.49	116.63	119.65	37.52	-1.52	0.93	36.97	-1.18	0.65	0.71
14	279	108.20	107.55	111.60	33.73	-0.88	0.42	33.38	-0.73	0.33	0.39
15	280	100.07	99.51	104.32	30.27	-0.81	-0.04	30.02	-0.97	0.01	0.31

continued on next page

	i	$R_i^{1/3}$	$G_i^{1/3}$	$B_i^{1/3}$	$L_i^{(c)}$	$a_i^{(c)}$	$b_i^{(c)}$	$L_i^{(t)}$	$a_i^{(t)}$	$b_i^{(t)}$	ΔE_i
\multicolumn{12}{l}{continued from previous page}
16	281	90.91	90.64	95.90	26.41	-0.91	-0.27	26.15	-1.03	-0.05	0.36
17	282	79.89	79.08	85.16	21.46	-0.44	-0.66	21.90	-0.87	0.03	0.92
18	283	69.12	68.01	75.05	16.68	-0.08	-1.25	17.48	-0.73	-0.27	1.42
19	284	60.49	59.22	67.08	12.87	0.16	-1.81	13.71	0.05	-0.82	1.30
20	285	54.43	54.15	61.01	10.46	-0.64	-1.08	10.70	-0.27	-0.85	0.50
21	286	50.36	49.97	56.43	8.59	-0.66	-0.73	8.91	-0.05	-1.10	0.78
22	287	47.74	47.31	53.31	7.38	-0.71	-0.34	7.28	0.48	-1.23	1.49
$D_{\max}$	288	42.24	42.22	48.45	5.08	-0.94	-0.58	3.47	1.17	-3.37	3.85

Some printer gamuts

We show here some views in the CIELAB space of the color gamuts of different printers, using different paper types and dithering techniques. The gamuts are illustrated using a basis of $5 \times 5 \times 5$ target colors regularly distributed in the printer RGB/CMY printer space, by using the color chart of Figure D.1, and the triangulation algorithm described in Chapter 5. The rendering of the gamuts is done using Matlab (Matlab Language Reference Manual, 1996).

In Figure D.2, we show the gamut of the Mitsubishi S340-10 sublimation printer. We see that the gamut is very regular. The CIELAB-values are measured using a SpectraScan spectrophotometer, and they are reported in Table D.1.

In Figure D.3, we show the gamut of the Epson Stylus 2 ink jet printer using coated paper. The dithering is regular, performed by a Ghostscript driver. The CIELAB-values are obtained using a scanner AGFA Arcus 2, calibrated by the methods described in Chapter 3. We see that this gamut is comparable to the sublimation gamut.

In Figure D.4, we show the gamut of the Epson Stylus 2 ink jet printer using glossy paper. Error diffusion dithering is performed using the printer driver for Windows delivered by Epson (a newer version). We note that the gamut is very distorted when compared to the previous ones. This is probably due to the fact that the printer driver incorporates a conversion from RGB monitor values to CMYK.

In Figure D.5, we show the gamut of the Epson Stylus 2 ink jet printer using normal paper. As expected, we note that the colors are much less saturated, compared to glossy paper. Another observation is that the black is far away from the other colors. It seems that this printer driver (the first version we used, the one that was delivered with the printer) adds black ink in an abrupt manner when approaching black.

In Figure D.6, we show the gamut of the Kodak sublimation printer. We see that the gamut is quite large, but that it has quite strange behavior when compared to the Mitsubishi gamut. It is very 'rounded'.

To conclude, we observe large differences in the sizes and shapes of the color gamuts of different printers, and also using the same printer with different paper and dithering methods.

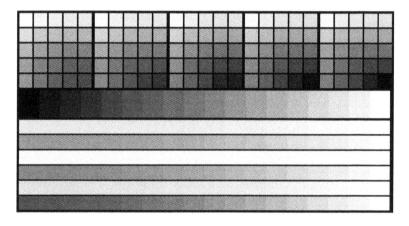

Figure D.1: *The regular* 5 × 5 × 5 *color chart used for the printer characterization described in Chapter 5, and for defining the color gamuts of this appendix. The lower part of the chart is mainly used for visual quality assessment.*

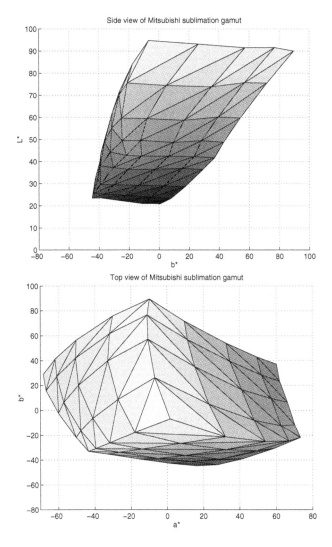

Figure D.2: *Two views of the color gamut of the Mitsubishi S340-10 sublimation printer.*

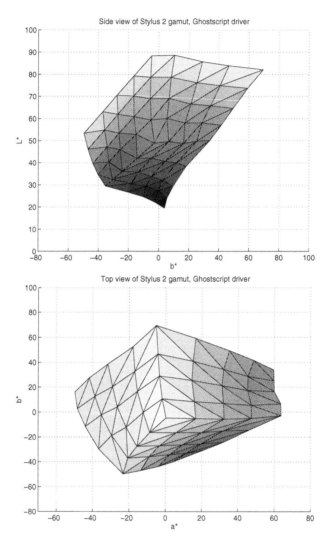

Figure D.3: *Two views of the color gamut of the Epson Stylus 2 ink jet printer using coated paper. The dithering is performed using a Ghostscript driver with regular dithering.*

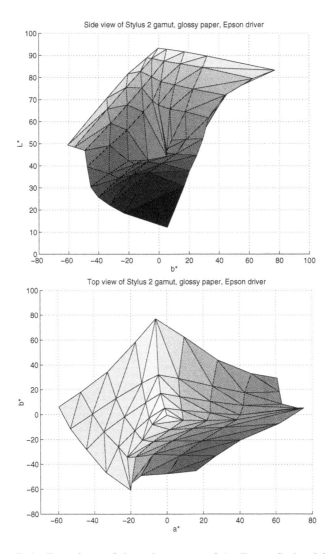

Figure D.4: *Two views of the color gamut of the Epson Stylus 2 ink jet printer used on glossy paper. The dithering is performed using the Epson printer driver for Windows.*

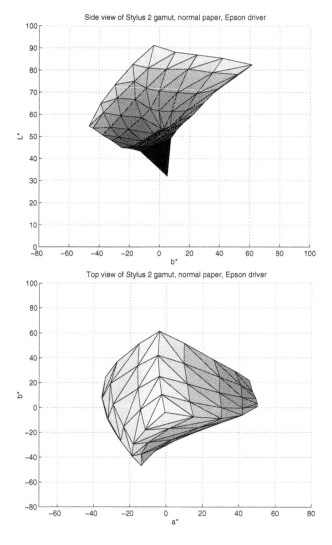

Figure D.5: *Two views of the color gamut of the Epson Stylus 2 ink jet printer using normal paper. The dithering is performed using the Epson printer driver for Windows.*

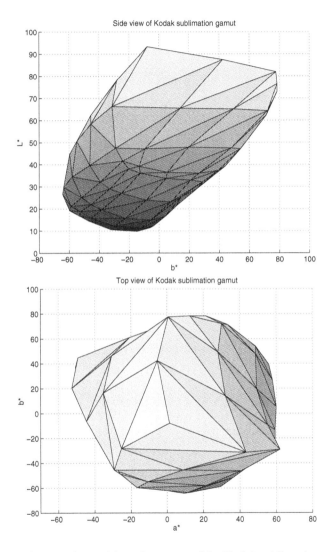

Figure D.6: *Two views of the color gamut of the Kodak sublimation printer*

Table D.1: *Printer RGB values and CIELAB values measured using a SpectraScan spectrophotometer for the 5 × 5 × 5 regular color chart printed on a Mitsubishi S340-10 sublimation printer*

R	G	B	L^*	a^*	b^*
255	255	255	94.7911	1.5955	-6.8684
192	255	255	84.0797	-17.8578	-17.6875
128	255	255	72.0422	-31.8078	-26.2133
64	255	255	62.9013	-39.4568	-30.7851
0	255	255	55.3688	-43.4886	-33.0856
255	192	255	75.5800	32.1785	-21.2899
192	192	255	70.1576	7.3172	-26.5871
128	192	255	62.1554	-13.9569	-31.1581
64	192	255	55.8448	-26.3830	-33.7420
0	192	255	50.0107	-32.9923	-35.1696
255	128	255	59.6239	53.9842	-25.1015
192	128	255	55.1744	32.4066	-31.6210
128	128	255	48.8970	9.7969	-36.2631
64	128	255	44.0945	-5.1532	-38.7098
0	128	255	40.8377	-15.3952	-39.1232
255	64	255	49.3922	66.8936	-24.6462
192	64	255	44.3251	49.2984	-32.4310
128	64	255	37.9678	29.1272	-39.2814
64	64	255	32.1464	12.9999	-40.9025
0	64	255	32.0358	1.6803	-42.6085
255	0	255	42.7946	73.3206	-21.8379
192	0	255	37.2602	59.0166	-30.4858
128	0	255	29.6293	41.5856	-38.9437
64	0	255	25.5686	27.6855	-43.7222
0	0	255	23.4675	16.1338	-44.5828
255	255	192	93.1684	-6.8833	26.2514
192	255	192	81.9985	-27.1124	8.6954
128	255	192	69.9304	-41.1033	-6.4384
64	255	192	60.7626	-47.6620	-16.2789
0	255	192	53.9492	-50.2212	-21.6511
255	192	192	74.4967	23.0797	4.9723
192	192	192	68.7936	-1.3803	-4.9399

continued on next page

continued from previous page					
R	G	B	L^*	a^*	b^*
128	192	192	61.1589	-21.7049	-15.3098
64	192	192	54.5825	-33.2980	-22.3473
0	192	192	48.9658	-38.7392	-26.2695
255	128	192	59.4418	47.9730	-7.9950
192	128	192	54.6585	26.0822	-15.7178
128	128	192	48.4687	3.8596	-23.8192
64	128	192	43.6237	-11.1695	-29.3715
0	128	192	40.1234	-20.2493	-31.7986
255	64	192	49.5749	62.6982	-12.2663
192	64	192	43.8511	44.7651	-20.5739
128	64	192	37.3261	24.1746	-28.8352
64	64	192	33.1185	8.1827	-34.2998
0	64	192	31.4640	-3.2578	-36.1161
255	0	192	42.4879	70.1752	-11.6463
192	0	192	36.4176	55.2515	-20.5704
128	0	192	29.5067	37.8333	-30.2856
64	0	192	25.2595	22.8294	-36.3859
0	0	192	23.5805	11.5306	-38.5804
255	255	128	91.4948	-10.6066	57.2467
192	255	128	80.6532	-32.7740	39.8274
128	255	128	68.3913	-48.9341	20.0761
64	255	128	59.3573	-56.9432	6.3887
0	255	128	52.9891	-59.9529	-2.3397
255	192	128	73.8660	15.2095	35.1096
192	192	128	68.1522	-9.4638	24.6715
128	192	128	59.7129	-30.6848	10.0268
64	192	128	53.5371	-43.4988	-0.5847
0	192	128	48.2964	-49.2802	-7.9236
255	128	128	58.6989	41.0806	16.6695
192	128	128	53.6376	18.3041	8.5068
128	128	128	46.7449	-4.6495	-2.6147
64	128	128	42.7194	-20.7832	-10.6433
0	128	128	39.0482	-29.6854	-16.0468
255	64	128	48.7261	57.2471	9.5510
192	64	128	43.0867	38.4412	0.0501
128	64	128	36.3744	16.8608	-10.8185
continued on next page					

continued from previous page

R	G	B	L^*	a^*	b^*
64	64	128	32.9135	-0.6972	-18.1295
0	64	128	30.5845	-11.7242	-22.3896
255	0	128	42.4778	66.1599	6.5373
192	0	128	36.5380	50.4223	-2.6073
128	0	128	29.2820	31.2173	-13.7199
64	0	128	24.7424	15.0922	-21.9196
0	0	128	22.3717	3.5016	-26.2323
255	255	64	91.5884	-11.0630	76.6991
192	255	64	82.1921	-31.7721	61.6557
128	255	64	69.4131	-50.4391	42.0293
64	255	64	59.9342	-60.8450	27.1295
0	255	64	51.8248	-66.8276	15.8290
255	192	64	75.7238	11.1285	56.9625
192	192	64	70.5586	-11.6472	48.2367
128	192	64	61.4111	-35.2094	33.8942
64	192	64	53.3225	-49.1434	20.5018
0	192	64	46.8250	-56.1747	10.5789
255	128	64	60.1234	34.5927	39.5181
192	128	64	55.7033	13.8508	31.5727
128	128	64	48.0872	-10.0487	19.2921
64	128	64	42.5174	-27.0552	9.1625
0	128	64	38.0642	-37.9667	1.7805
255	64	64	49.4438	51.9812	28.2829
192	64	64	44.2822	34.3503	20.4524
128	64	64	36.7970	11.0085	8.7455
64	64	64	32.1028	-6.7321	-0.3339
0	64	64	28.9294	-19.3686	-6.6491
255	0	64	42.2257	61.9382	23.3849
192	0	64	36.8812	47.0011	15.3282
128	0	64	28.9768	26.3447	3.3116
64	0	64	23.9661	8.7011	-5.3829
0	0	64	21.0778	-3.8731	-11.4199
255	255	0	90.0119	-9.7284	89.5759
192	255	0	80.9503	-29.9992	75.8667
128	255	0	67.7903	-50.1668	56.4221
64	255	0	58.1977	-61.8021	41.2901

continued on next page

| continued from previous page | | | | | |
R	G	B	L^*	a^*	b^*
0	255	0	50.7141	-68.3713	28.8505
255	192	0	75.5604	9.5194	71.6834
192	192	0	70.0461	-12.0482	63.1327
128	192	0	60.5593	-36.1611	48.6978
64	192	0	52.7365	-50.9991	35.7617
0	192	0	46.3704	-59.5632	24.3783
255	128	0	59.7103	32.5537	54.0855
192	128	0	55.1988	12.8982	47.1941
128	128	0	47.5119	-11.3343	35.0374
64	128	0	41.9822	-30.1078	24.8570
0	128	0	37.3402	-42.3133	15.9954
255	64	0	48.7629	49.4398	42.9982
192	64	0	43.8249	32.7111	35.9934
128	64	0	36.5525	9.0782	24.3247
64	64	0	31.6048	-10.3024	14.7665
0	64	0	28.4563	-24.0073	7.1881
255	0	0	41.6645	60.2547	37.0043
192	0	0	36.4131	45.2382	29.1444
128	0	0	28.6380	24.6906	17.2451
64	0	0	23.3157	5.9978	7.7766
0	0	0	20.9685	-7.3417	0.7809

Appendix **E**

Gamut mapping techniques

The color gamut of a device such as a printer is defined as the range of colors that can be reproduced with this device. Gamut mapping is needed whenever two imaging devices do not have coincident color gamuts, in particular when a given color in the original document cannot be reproduced with the printer that is used. Several researchers have addressed this problem, see for example the following references: (Stone *et al.*, 1988, Gentile *et al.*, 1990, Stone and Wallace, 1991, Pariser, 1991, Hoshino and Berns, 1993, MacDonald, 1993b, Wolski *et al.*, 1994, Spaulding *et al.*, 1995, MacDonald and Morovič, 1995, Katoh and Ito, 1996, Luo and Morovič, 1996, Montag and Fairchild, 1997, Tsumura *et al.*, 1997, Morovic and Luo, 1997; 1998).

We present here a resume of the gamut mapping techniques most frequently reported in the literature. First different *gamut clipping* techniques are presented. Gamut clipping is absolutely necessary in any image reproduction system, to assign a reproducible in-gamut color to any out-of-gamut color. In-gamut colors are not modified. Then we present different continuous gamut mapping transformations that modifies all colors of the image, both in-gamut and out-of gamut colors. These transformations are intended to reduce the unwanted effects of gamut clipping, by assuring smooth and continuous color changes.

1. **Gamut clipping.** This is the basic gamut mapping technique that consists in clipping out-of-gamut colors to a color on the gamut

boundary. Colors that are inside the gamut are not changed. Different strategies might be employed, as illustrated in Figure E.1.

(a) **Orthogonal clipping.** This is the clipping that induces the smallest perceptual ΔE_{ab} error. However it might give unwanted hue changes.

(b) **Constant-luminance clipping.** Out-of-gamut colors are clipped to the nearest boundary color with the same hue and luminance. That is, only the saturation is changed. However, this method induces problems for colors with luminance that exceeds the minimum or maximum luminance of the reproduction gamut.

(c) **Radial clipping.** Out-of-gamut colors are clipped to the nearest gamut boundary color in the direction towards the mid-gamut point $L^* = 50, a^* = b^* = 0$.

(d) **Truncating in output color space.** This clipping method consists in simply truncating the color coordinates in the output (CMY/RGB printer) color space. F.ex. if the result of a transformation from CIELAB to CMY results in

$$[C, M, Y] = [-10\%, 50\%, 110\%].$$

this would be truncated to

$$[C, M, Y] = [0\%, 50\%, 100\%].$$

This method does not provide any control of visual color difference, as opposed to the previously mentioned methods.

2. **Gamut compression.** It is generally regarded as more important to preserve the relative color differences between all colors, rather than to preserve the absolute coordinates of those in gamut.(MacDonald and Morovič, 1995) To obtain this, gamut compression can be applied to all colors of an image by radial scaling of all coordinates in CIELAB towards a mid-gamut point, as shown in Figure E.2. The compression factor might be determined dependent on the image such that all colors of the image is compressed to be in gamut, or chosen as a standard value, in which case care must be taken to clip properly colors that are still out of gamut after compression, such as $\mathbf{C}'_1$ in Figure E.2.

3. **Lightness compression.** This is a one-dimensional mapping along the L^* axis such that the range of luminances in the original image is mapped onto the range of luminances in the reproduction gamut. This method is also called tonal mapping.

4. **White point adaption.** This technique accounts for the color of the printing paper by applying a geometrical translation or deformation of the colors of the original image such that the white of the original image, or eventually the white point of the input technology, is reproduced as clean paper with no ink. This corresponds to *relative colorimetry* in the ICC terminology (ICC.1:1998.9, 1998).

All these gamut mapping techniques can be combined to find the best compromize to solve the gamut mismatch problem.

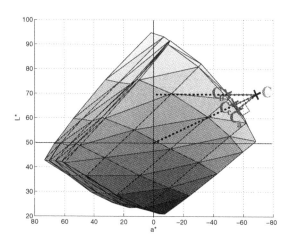

Figure E.1: *Gamut clipping methods.* **C** *Represents the original out-of-gamut color,* $\mathbf{C}_a$ *the color after orthogonal clipping,* $\mathbf{C}_b$ *after constant-luminance clipping, and* $\mathbf{C}_c$ *after radial clipping.*

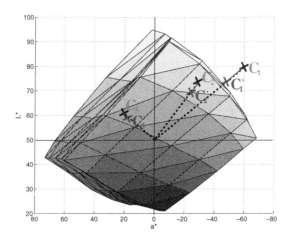

Figure E.2: *Gamut compression.* $\mathbf{C}_1$, $\mathbf{C}_2$ *and* $\mathbf{C}_3$ *are mapped towards the gamut center, resulting in* $\mathbf{C}_1'$, $\mathbf{C}_2'$ *and* $\mathbf{C}_3'$. *Note that* $\mathbf{C}_1'$ *is still out of gamut after compression.*

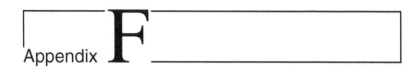

Appendix F

Bibliography on the dimensionality of spectral reflectances

How many components are needed to represent the spectral reflectance of a surface? What is the dimension of a reflectance spectrum? how many channels are needed for the acquisition of a spectral reflectance? Such questions have been discussed extensively in the literature. We have done a survey of the literature concerning this question. See Section 6.4 for our analysis of this subject.

Munsell spectra

▓ Possibly the first attempt to fit a linear model to a set of empirical surface spectral reflectances was performed by Joseph Cohen (1964) of the University of Illinois back in 1964. He analyzes a subset of 150 out of 433 Munsell chips, and concludes that they depend on only three components, which account for 99.18% of the variance. (92.72, 97.25, 99.18, 99.68). He do not, however evaluate spectral reconstruction from these 3 values. Furthermore it may be noted that Cohen's analysis is applied to a quite small subset of the Munsell colors.

■ Eem *et al.* (1994) of Kum Oh National University of Technology and Dae Jin University in Korea find that the spectral reflectance functions of the Macbeth Color Checker can be reconstructed very closely by the first four characteristic vectors calculated by PCA from the reflectance spectra of 1565 glossy Munsell chips. The first eight characteristic vectors and eigenvalues (5.1775, 0.3695, 0.1193, 0.0204, 0.0091, 0.0050, 0.0031, 0.0022) are reported. The original and reconstructed spectra are represented in the CIE chromaticity diagram, and no quantitative data is given for the errors.

■ Maloney (1986) of the University of Michigan applies the same analysis as Cohen to a more complete set of 462 Munsell surface spectral reflectances, as well as to a set of natural spectra measured by Krinov (1947). The proportions of variance accounted for by a linear model with 2-6 parameters are given to be 0.9583, 0.9916, 0.9959, 0.9985 and 0.9992. The linear model based on Munsell data is found to be quite appropriate also when applied to the Krinov data. He concludes that a linear model of five to seven parameters is appropriate.

■ Burns and Berns (1996) of the Munsell Color Science Laboratory at the Rochester Institute of Technology, NY present a seven-channel camera for multispectral image capture using a set of seven interference filters manufactured by Melles Griot, sampling the visible wavelength range at intervals of approximately 50nm. They evaluate three methods of spectral reconstruction from the seven camera signals: cubic spline interpolation, modified-discrete-sine-transform (MDST) interpolation (Keusen, 1994, Keusen and Praefcke, 1995), and principal component analysis (PCA). The PCA is applied using Munsell colors, apparently using only the first five characteristic vectors. The 7 camera responses are mapped onto the 5-dimensional spectral representation using a LMS approach.

The procedures are evaluated using a MacBeth Color Checker chart, with ΔE_{ab} as error measure, calculated from the spectra using CIE Illuminant A and the 2° observer, using 10nm spectrophotometric measurements as reference. A direct mapping from the seven camera signals to CIELAB using a first order least-square model is also evaluated. The PCA method outperforms the rest; giving mean and max ΔE_{ab} errors of 2.2 and 4.7.

In (Burns, 1997) it is stated that at least six basis vectors are needed for critical applications where an average error of $\Delta E_{ab} \leq 1.0$ is needed.

■ In (Jaaskelainen *et al.*, 1990), Jaaskelainen, Parkkinen, and Toyooka performs an analysis of the Munsell chips reported earlier (Parkkinen *et al.*, 1989) together with a set of 218 naturally occurring spectral reflectances, to form two linear bases by PCA. They find that the basis determined using the Munsell spectra can be used to represent the natural spectra. However, more components are needed to attain the same accuracy.

■ In a recent study, Lenz and Österberg of Linköping University together with the fins Hiltunen, Jaaskelainen and Parkkinen (Lenz *et al.*, 1995; 1996a) investigate three databases of spectral reflectances, the Munsell colors of Parkkinen *et al.* (1989), a new set of 1269 Munsell colors measured by a more accurate (1nm) spectrophotometer, and a set of 1513 spectra based on the NCS (Hård *et al.*, 1996) color system. They find by PCA that the databases have very similar statistical properties, and that the first few eigenvectors developed from the different databases are highly correlated. They further present a class of systems that find a set of positive basis vectors for these spaces with only slightly higher reconstruction errors compared to the PCA basis. No conclusion is drawn on the number of basis spectra needed, but 6 is used as an example, giving a reconstruction error of about 4%.

■ Parkkinen *et al.* (1989) measured and analyzed by PCA/K-L a set of 1257 reflectance spectra of the Munsell chips measured by a acusto-optic spectrophotometer. Contrary to both Cohen's (Cohen, 1964) and Maloney's (Maloney, 1986) previous analyses, it was found that as many as eight eigenvectors were necessary, giving mean and max spectral reconstruction errors of 0.008 and 0.02, respectively. Interesting graphical representations of the errors presented include the error bands, that is, the area between the maximum positive and maximum negative wavelenght-wise reconstruction error. The cumulative information content is not reported, but it may be calculated approximately from the eigenvalues of the correlation matrix, (1129.2, 72.7, 28.8, 12.7, 5.0, 3.4, 2.2, 0.8), giving (0.900, 0.958, 0.981, 0.991, 0.995, 0.998, 0.999, 1)[1].

---

[1] Remark that this is surely an overestimation of the cumulative information content, since

▓ Wang *et al.* (1997) analyzes the database of 1269 Munsell spectra mentioned above (Parkkinen *et al.*, 1989). By a neural network approach they design a set of 8 filters having strictly positive spectral reflectances. The space spanned by these filters is found (by visual comparison) to be quite close to the space spanned by the first 8 optimal basis vectors calculated by Karhunen-Loeve expansion (PCA). The mean and maximal spectral reconstruction error using these 8 filters is found to be 0.07% and 2.6%, respectively.

See also (Hauta-Kasari *et al.*, 1998) which is more complete than the above.

Natural reflectances

▓ Dannemiller (1992) of the University of Wisconsin makes an attempt to answer the question of how many basis functions that are necessary to represent the spectral reflectance of natural objects. His analysis is applied to the set of 337 spectral reflectance functions measured by Krinov (Krinov, 1947). He realizes that statistical measures based only on the spectral data might not be appropriate to evaluate the quality of the approximation of a spectrum by a reduced set of basis functions. He propose to apply an *ideal observer analysis* with which an ideal observer was placed at the level of photon catch in the foveal photoreceptors of a typical human eye. The performance is evaluated using a measure of visual color matching d' based on the number of photons absorbed by the cones. Based on this measure d', or rather on its rate of decrease of d', as the absolute value depends on irrelevant factors such as the level of illumination etc., he comes to the conclusion that three PCA eigenvectors are necessary and probably sufficient for representing the spectral reflectance functions of natural objects.

In his analysis, an interesting illustration is presented, namely the frequency distribution showing the number of basis functions required in an approximation to produce a given d', this revealing a bimodality of his data set. A further analysis suggests that inanimate materials have simpler reflectance spectra than animate materials.

---

the variance of the eigenvectors beyond 8 is not taken into account.

■ Vrhel *et al.* (1994) present a new set of 354 spectral reflectances of an ensemble of different materials, including 64 Munsell chips, 120 Du Pont paint chips, and 170 reflectance spectra from various natural and man-made objects. These data are proposed as a replacement of the data measured in 1947 by Krinov (1947), as this dataset is somewhat limited. A principal component analysis is performed using these data to create the covariance matrix. They report the reconstruction error using from 3 to 7 basis functions determined by PCA. The errors are measured as average and maximum ΔE_{ab} and square spectral error. The square errors in the CIE xy chromaticity diagram are also reported graphically.

We note that even with seven basis functions, the maximum $\Delta E_{ab} = 5.05$. The cumulative information content is not reported, but may be calculated from the spectra found at ftp://ftp.eos.ncsu.edu/pub/spectra/ to be: (0.4471, 0.6810, 0.8053, 0.8536, 0.8874, 0.9110, 0.9304, 0.9452, 0.9567, 0.9661, 0.9729, 0.9782, 0.9828, 0.9866, 0.9893, 0.9915). We note that as much as 16 basis vectors have to be taken into account if 99% of the information content is to be preserved.

■ Praefcke and Keusen (1995) of Aachen University of Technology propose to represent reflectance spectra using a set of basis function optimized to minimize the mean or maximum ΔE_{ab} errors under a set of different illuminants. Two different nonlinear stochastic optimization algorithms are evaluated, and the results are compared to those obtained by PCA. The analysis is applied on the dataset as published by Vrhel *et al.* (1994). They obtain generally smaller errors when compared to the PCA approach. Note however that as a result of choosing the mean or maximum ΔE_{ab} error as optimization criterion, the other error measure often turns out to be greater compared to PCA. It is concluded that five basis vectors seems to be appropriate.

This analysis is pursued in (Praefcke, 1996) where it is noted that the basis functions optimized as described above do have a high degree of cragginess or roughness. This is not desirable, and Praefcke proposes thus a solution where the criterion to be minimized comprises both ΔE_{ab} and a measure of roughness. The resulting basis functions show only slightly lower performance than the optimal ones.

▓ Keusen and Praefcke (1995), Keusen (1996) of Aachen University of Technology propose a multispectral color system with an encoding format compatible with the conventional tristimulus model. Twelve (Keusen and Praefcke, 1995) or fourteen (Keusen, 1996) interference filters are used for acquisition, giving reconstruction errors lower than $\Delta E_{ab} = 1$. They evaluate the reconstruction error of 354 spectra measured by Vrhel *et al.* (1994) in terms of ΔE_{ab} under different illuminants, using their basis spectra compared to basis spectra issued from a PCA analysis, and from 3 to 7 components. When defining that the maximal error should be less than $\Delta E_{ab} = 3$, seven components are needed when only natural illuminants are used, but up to 10 when illuminants such as F2 is introduced.

Human skin reflectances

▓ In an analysis by Imai *et al.* (1996b;a) from the group of Miyake at the university of Chiba, a set of 108 reflectance spectra of skin in faces of 54 Japanese woman are analyzed by PCA. The cumulative contribution ratios of the PCs are presented, and it is noted that the three first components represent 99.5% of the signal. They proceed thus to an estimation of spectral reflectance from tristimulus values, and from a 3-channel HDTV camera.

Painting reflectances

▓ Tsumura, Miyake and others from the university of Chiba (Haneishi *et al.*, 1997, Yokoyama *et al.*, 1997) perform a PCA on a set of 147 oil paint samples. They present the cumulative contribution ratio of the principal components, and conclude that the spectral reflectance of the paintings can be estimated 99.32% by using a linear combination of 5 principal components. (Remark that it is not completely clear from (Yokoyama *et al.*, 1997) if the analysis is based merely on the 147 oil paint samples, or on 'several thousands color patches') They decide thus to acquire spectral images using a CCD camera with 5 filters. They present a simulated annealing method to design five optimal Gaussian filters, giving a mean and max ΔE_{uv} error applied to the 147 oil paint samples of 0.22 and 0.63, respectively. They also present an optimal choice of 5 filters from a set of

24 available filters by an exhaustive search, giving an average color difference of $\Delta E_{uv} = 1.16$.

▓ Maître *et al.* (1996) of the Ecole Nationale Supérieure des Télé-communications in Paris present a method for the reconstruction of the spectral reflectance function of every pixel of a fine-art painting, from a series of acquisitions made through commercially available chromatic filters. Procedures for the choice of filters is presented. It is stated that the 12 largest eigen-vectors represent 98.2% of the global energy, and that the use of 10 to 12 acquisition filters is necessary for adequate spectral reconstruction.

▓ García-Beltrán *et al.* (1998) of the University of Granada, Spain, performs an analysis of a set of 5574 samples of acrylic paint on paper. The samples were generated by a mixture of 24 basic commercial acrylic colors for artists. They found that the first seven vectors of the linear basis were sufficient for a more than adequate mathematical representation of the spectral-reflectance curves. They use a goodness-fitting coefficient (GFC) to evaluate the quality of fit. They also perform another analysis dividing the data into five hue groups, red, yellow, green, blue, and purple. By doing this they reduce the number of vectors needed.

Other spectra

▓ Young (1986) found that the first three basis spectra accounted for 93% of the variability in a set of spectra consisting of 441 twelve-component spectra of macaque lateral geniculate nucleus.

▓ In an approach to classification of optical filters, the transmission spectra of a set of 23 blue transparent filters was analyzed by Karhunen-Loeve (PCA) analysis by Parkkinen and Jaaskelainen (1987) of the University of Kuopio, Finland. The cumulative information content for the first ten eigenvectors is reported (79.5, 89.7, 94.1, 98.0, 99.0, 99.5, 99.8, 99.9, 100.0, 100.0) along with the reconstruction error presented in a figure. The reconstruction error decreases rapidly, from approximately 5.5% using only one eigenvector, via 3% using three and 1% using six eigenvectors. Another interesting measure that is presented graphically is the minimum length from the projections to the subspace formed from the sample set. This increases quite rapidly towards a value of one. For their classification purpose, the representation of the spectral transmittances using three eigenvectors was found to be appropriate.

■ Sato *et al.* (1997) from Toyohashi University of Technology, Japan, estimates the spectral reflectance from RGB values using PCA and a neural network. They analyse a set of spectral reflectances of 1803 JIS (Japan Industrial Standard) color chips. They analyse the dimensionality of the data by PCA. Using three parameters, 99.45% of the signal variance is accounted for, and the mean and maximal reconstruction error expressed in ΔE_{ab} under D_{50} illuminant is of 2.92 and 21.67, respectively. The corresponding numbers using four parameters is 0.9989, 1.01 and 9.59. Based on these results, they find that a representation using the first four principal component vectors is appropriate. In a practical experiment with RGB acquisition under unknown illuminant the accuracy of conversion from RGB to 4-dimensional surface reflection using a trained NN is reported to be better than 98.5%, giving mean color differences between measured and estimated spectral reflectances ranging from about $\Delta E_{ab} = 4.74$ to 7.17.

In an earlier study by Arai *et al.* (1996), their NN color correction method is compared to two classical methods based on PCA and white point mapping. The NN method outperforms the other methods. To explain this result, they compare the spectral reconstruction error using 3 and 4 principal components with their NN approach. The MSE (root?) on 1115 samples printed by a dye sublimation printer is found to be 0.123% using 3 PCs, 0.0691% using 4, and 0.0364 using NN.

Realization of multispectral acquisition systems

For a practical realization of a multispectral scanner using filters, the dimensionality of reflectances is highly relevant when designing the system, in particular when choosing the number of acquisition channels. Also here, the choices found in the literature are many:

■ Imai *et al.* (1996b;a) uses only three channels, see above.

■ Another analysis by Shiobara *et al.* (1995; 1996) from the same research group is applied to a set of 310 spectral reflectances of the gastric mucuos membrane. The cumulative contribution ratio is presented, and also here the three first components represent 99.5% of the signal. Given the actual spectral characteristics of an electronic endoscope camera, including the three color filters, a spectral reconstruction is simulated on the 310 samples, giving mean and max ΔE_{uv} reconstruction errors of 2.66 and 9.14, respectively. A practical realization is done.

■ Chen and Trussell (1995) of North Carolina State University presents an approach to designing color filters for a colorimeter that uses multiple internal illuminants and multiple filtered detectors. A set of four filters was used, giving an average ΔE_{ab} of 2.3 for photometric measurements. The correlation matrix was obtained from a Dupont reflectance data set.

■ Haneishi *et al.* (1997), Yokoyama *et al.* (1997) use 5 filters, see above.

■ Kollarits and Gibbon Kollarits and Gibbon (1992) tested the use of five filters for a television application, achieving significant improvement in color errors compared to a typical three-filter camera.

■ Tominaga (1996; 1997) of the Osaka Electro-Communication University presents a multichannel vision system comprising six color channels, designed to recover both the surface spectral reflectance and the illuminant spectral power distribution from the image data. Six bandpass Kodak Wratten filters, noted B, BG, G, Y, R and R2, having peak transmission values spread over the visible spectrum are used. The choice of 6 channels were motivated from literature studies (Maloney, 1986, Parkkinen *et al.*, 1989, Vrhel *et al.*, 1994) where it is stated that spectral reflectances of natural and artificial objects may be represented using five to seven basis functions. However, when modeling spectral reflectance, Tominaga finds that a dimension of 5 is appropriate. The average reflectance estimation error in an experiment using two scenes containing three cylinders of colored plastic and paper is reported to be 0.023 and 0.032, respectively. Note that this systems aims for color constancy, in that it estimates the illuminant as well, by taking into account specular reflections in the scene.

- Saunders, Cupitt and Martinez (Saunders and Cupitt, 1993, Martinez *et al.*, 1993) at the National Gallery of London presents an image acquisition system using a set of seven broad-band nearly Gaussian filters. The filters have peak transmittances ranging from 400 to 700 nm in steps of 50 nm, and a half-height bandwidth of about 70 nm.

- Burns and Berns (1996) use seven filters, see previous description.

- Abrardo *et al.* (1996) at the University of Florence, Italy, presents a multispectral scanner using 7 color filters. Using this scanner, and a color calibration, they obtain a color accuracy of $\Delta E_{ab} = 2.9$, evaluated on a subset of 20 patches of the AGFA IT8.7/3 color target.

- Vrhel and Trussell (1994) evaluates the colorimetric quality of an acquisition system using from 3 to 7 optimized filters. A practical realization is presented in (Vrhel *et al.*, 1995).

- Hardeberg *et al.* (1998a; 1999) propose five to ten channels.

- Maître *et al.* (1996) propose ten to twelve channels.

www.ingramcontent.com/pod-product-compliance
Lightning Source LLC
Chambersburg PA
CBHW051045050326
40690CB00006B/601